CINQ TRAITÉS D'ALCHIMIE

COLLECTION OF RELATED WORKS
ON THE HERMETIC SCIENCES

FIVE TREATIES ON ALCHEMY

FROM THE GREATEST PHILOSOPHERS
PARACELSUS, ALBERT THE GREAT, ROGER BACON, R. LULLE,
ARNAUDE VILLENEUVE TRANSLATED FROM FRENCH INTO LATIN
By *Christopher Templesage*

PRECEDED BY
THE EMERALD TABLET, FOLLOWED BY A GLOSSARY

POST LABOREM SCIENTIAM.

~ Send ~
~ Messages & Notes To~
~Institutum Lapidarium Philosophorum~
ILP
PO BOX 994
Merlin, Oregon
97532
U.S.A.

TABLE OF CONTENTS

Preface……………………………………….6

Emerald Table……………………………….7

Biographical note on Arnauld de Villeneuve ….21

Le Chemin du Chemin………………………23

Biographical note on R. Lulle………………..30

The Clavicle………………………………… 31

Biographical note on Roger Bacon……………49

The Mirror of Alchemy………………………50

Biographical note on Paracelsus………………65

The Treasury of Treasures……………………67

Biographical note on Albert the Great…………78

The Compound of Compounds………………79

Glossary………………………………………105

(Z) Store the product in a suitable container and label it appropriately.

PREFACE

The current sciences are the daughters of mysterious sciences whose origin is lost in the mists of time, alchemy is the mother of chemistry, astrology preceded astronomy, at the base of mathematics we find the cabala and qualitative geometry, in principle history merges with mythology, medicine was taught to men by a god.

One knows a science well only when one knows its history. From the mother idea which founded science until nowadays, what incessant efforts, what trial and error! We take advantage of the work of our predecessors, carelessly, without thinking of the enormous amount of physical and intellectual work they have expended to pave the way for us. Many have spent their lives, spent their fortune, renounced pleasures and honors for love science.

How many have died as martyrs claiming to the last breath the eternal truth! It is Roger Bacon, persecuted all his life by ignorant monks, it is the learned Hypatia stoned by the populace of Alexandria, it is Averroës thrown into prison then exiled, for having advanced ideas contrary to the Koran, it is Bernard le Trévisan hated and tormented by his furious parents to see him spend his fortune in alchemical research, it is Denis Zachaire assassinated by his cousin to whom he had refused to reveal the secret of the philosopher's stone, it is Cardan , poor all his living and dying of grief, these are Perrot and Paracelsus, finishing their career on a hospital bed, they are Bernard Palissy and Borri dead in prison.

To do justice to these great men by bringing their work to light, by bringing them to life in their works, such has been our goal. However, their works have become rare, the large libraries alone could provide researchers with sufficient documents, but we know how difficult it is to obtain permission to work in a public library. On the other hand to form a private collection is very expensive and requires time and patience, often one does not finds that after several years of research the book that one desires; finally, most of these treatises are written in barbaric Latin, in an obscure style that is very tiring to read. All these reasons have led us to publish these translations. The authors have been carefully chosen among the most great names in alchemy:

Arnauld de Villeneuve, Raymond Lully the enlightened doctor, Albert the Great,

embracing all in his vast erudition, Roger Bacon the admirable doctor, anticipating his century and substituting experience and observation to the hollow ramblings of scholastics, finally Paracelsus, the great Paracelsus, upsetting the old theories, combining alchemy with medicine, never did a man have a greater influence on his century. We took the most important treaties, four out of five are translated for the first time into French. As for the translation is as exact as possible, the obscure passages are rendered word for word; we are attached to giving the sentence the turn it has in the text. Finally, the treatises are preceded by a biographical note and a bibliographical index.

We end with a piece of advice: to read this book without being prepared for it is to expose yourself to not understanding it, so you should read well beforehand: "Alchemy and the Alchemists" by M. Louis Figuier or : "The Origins of Alchemy" by M. Berthelot. For people who do not have time to read these two books, here is a quick words what Alchemy is: "It is, says Pernety, the art of working with nature on bodies to perfect them. The main purpose of this science is the preparation of a compound: the philosopher's stone, having the property of transmuting molten metals into gold or silver. The raw material of the philosopher's stone is the Mercury of the philosophers. It is given the property of transmuting by subjecting it to various operations, during which it changes color three times: from black, it becomes white, then red. White, it constitutes the white elixir or small stone, which changes metals into silver. Red, it constitutes the medicine or red elixir or great stone which changes metals into gold.

Alb. Poisson [a.k.a. A.. Fish. or White Fish]

FIVE TREATIES ON ALCHEMY
OF THE GREATEST PHILOSOPHERS
PARACELSE, ALBERT LE GRAND, ROGER BACON, R. LULLE, ARN. DE VILLENEUVE

NOTICE ON THE EMERALD TABLET OF HERMES

The Emerald Table of Hermes Trismegistus, (Le' Thaut) of Egyptian origin is the cornerstone of alchemy. Philosophers cite it all the time, so does it matter to know this document. It is found in all the important collections of hermetic treatises: Theatrum Chemicuin, Bibliotheca Chemica Mangeli, Bibliotheca contracta Albinei, Library of Alchemical Philosophers by Salmon, etc. The following translation is that of the Library of the alchemical philosophers of Salmon revised and corrected according to the Latin text which is at the head of the Bibliotheca chemica contracted Albinei.

EMERALD TABLET

It is true, without lie, certain and very true.
What is below is like what is above
and what is above is like what is below,
to perform the miracles of one thing. And of
just as all things came out of a thing by the thought of One, so all things arose from this thing by adaptation.
His father is the Sun, his mother is the Moon, the wind has
carried in her womb; the earth is its nurse.
It's here the father of all Thelêm.e of the Universe. His power is boundless on earth.
You will separate the earth from the fire, the subtle from the gross, gently, with great industry.
He ascends from earth to heaven, and immediately descends again to earth, and he gathers the force of things superior and inferior.
You will thus have all the glory in the world, therefore all darkness will depart from you.
It is the strong force of all force, for it will overcome all subtle things and penetrate all solid things.

It is how the world was created. Here is the source of admirable adaptations indicated here. This is why I was called Hermes Trismegistus, possessing the three parts of Universal Philosophy. What I have said of the operation of the sun is complete.

Visita
Interiora
Terræ
Rectificando
Inuenies
Occultum
Lapidem

CINQ TRAITÉS D'ALCHIMIE 12

Visita
Interiora
Rectificando
Invenies
Occultum
Lapidem

Visita
Interiora
Terre
Rectificando
Inuenies
Occultum
Lapidem

CINQ TRAITÉS D'ALCHIMIE

CINQ TRAITÉS D'ALCHIMIE

ARNOLDI DE VILLANOVA
SEMITA SEMITÆ

THE PATH OF THE PATH
by ARNAULD de VILLENEUVE

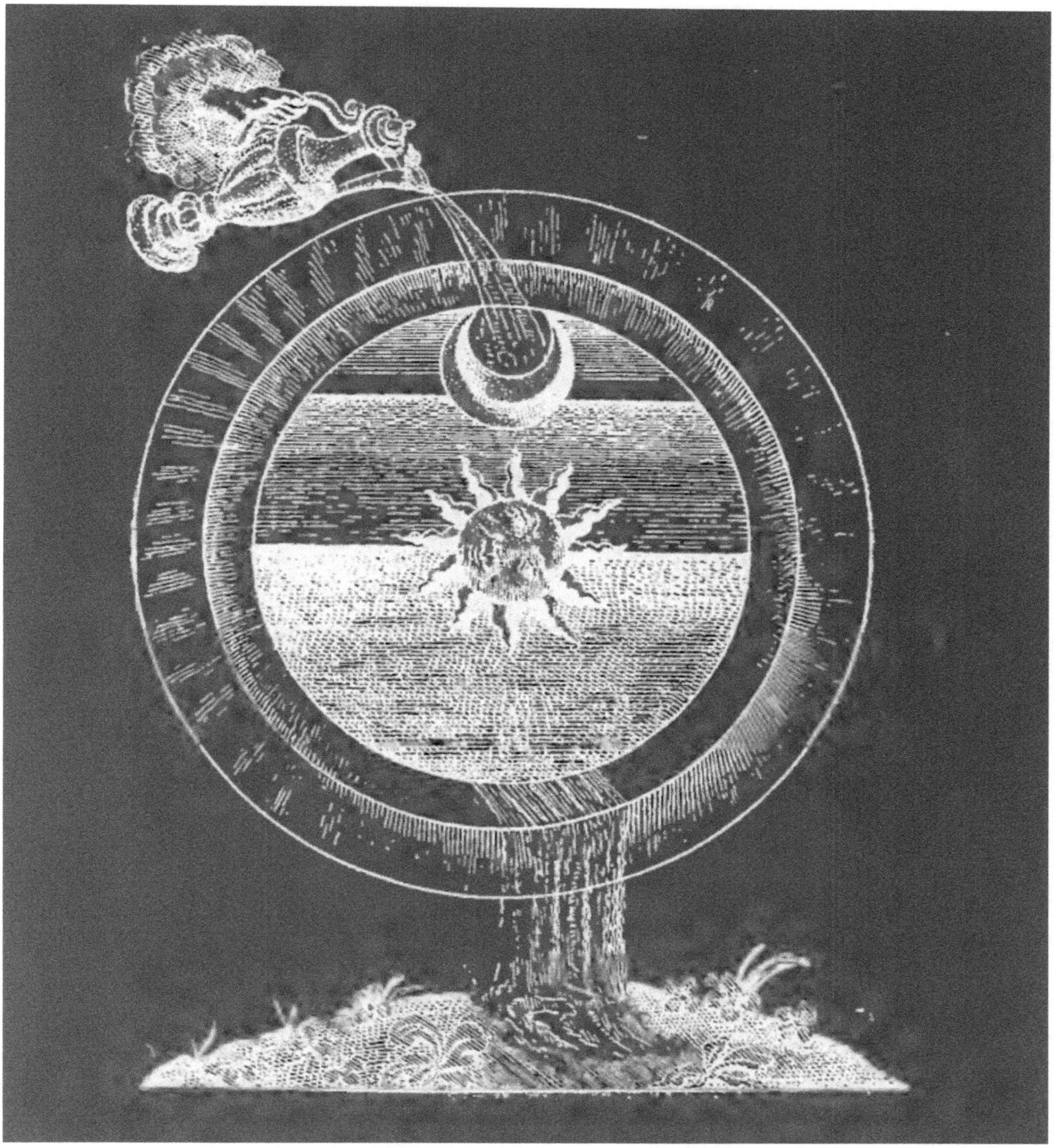

BIOGRAPHICAL NOTE ON ARNAULD DE VILLENEUVE

Arnauld de Villeneuve was born around 1245 France, as attested by Symphorianus Campegius and Joseph by Haitze. As for the precise place of his birth, it is uncertain. He studied dead languages in Aix, medicine Montpellier. He came to Paris to perfect himself; popular rumor accusing him of necromancy and alchemy, he fled to Montpellier, where he was soon appointed professor, then regent. In 1755 still showed in Montpellier, his house bearing carved on the facade a lion and a snake biting its tail. The thirst to learn makes it pass in Spain, it professes some time alchemy in Barcelona (1286) and learns Arabic.

He then visits the famous universities of Italy: Bologna, Palermo, Florence. He returns to Paris, but his heretical proposals, having excited theologians against him, he cautiously fled to Sicily, where Frederick II took him under his protection. Pope Clement V reaches the stone, summoned Arnauld de Villeneuve to him, with a promise of forgiveness. Arnauld embarked for France (the popes then sat in Avignon).

resolve into Mercury; therefore this Mercury is the prime matter of all metals. I will teach more away the way to do this transmutation, destroying thus the opinion of those who claim that the form of metals cannot be changed. They would be right if metals could not be reduced to their raw material, but I will show that this reduction in the matter first is easy and that transmutation is possible and feasible. For everything that is born, everything that grows, multiplies after its kind, like trees, people, the herbs. One seed can produce a thousand other seeds. So it is possible to multiply things ad infinitum. From what precedes, the one who analyzes things will see that if the Philosophers spoke in a way obscure, at least they told the truth.

They said in effect that our Stone has a soul, a body and a spirit, which is true. They compared his imperfect body to the body, because it is without power by itself; they called Water a vital spirit, because it gives the body, imperfect in itself and inert, the life which it did not have before and that it perfects its form. They called the ferment soul, because as we will see later, it also gave life to the imperfect body, it perfects it. and changes it into its own nature.

The philosopher says; "Change natures and you will find what you are looking for. " This is true. Because in our magisterium we first draw the subtle from the thick, the spirit of the body, and finally the dry of the wet, that is to say the land of Water, this is how we change natures; what was below we put above, so that the spirit becomes body, then the body becomes spirit. Philosophers still say that we made our Stone of a single thing and with a single vessel; And they are right. All our magisterium is drawn from our Water and is done with it. It dissolves metals themselves, but not by changing into the water of the cloud, as the ignorant believe. She burns and reduced to earth. She turns bodies to ashes, it incinerates, bleaches and cleans, according to what Morien says;

"The Azoth and the fire cleanse the Brass, that is to say wash it and completely remove its blackness. "Brass is an impure body, azoth is quicksilver.

Our Water unites different bodies between them, if they have been prepared as it has just been said; this union is such that neither fire nor any other force can separate them by combustion from their igneous principle. This transmutation steals the bodies, but that is not the vulgar sublimation of simple-minded, inexperienced people, for whom to sublimate is to elevate. These people take calcined bodies, mix them with sublimable spirits, that is to say mercury, arsenic, sulfur etc., and they sublimate the whole with the help of a high heat.

The charred bodies are carried away by the spirits and they say they are sublimated. But what is their disappointment, when they find impure bodies with their spirits more impure than before! Our sublation does not consist in elevating; the sublimation of the Philosophers is an operation which makes something vile and corrupted (by the earth) something purer. Similarly when we say commonly; So-and-so has been raised to the Episcopate... by "raised" is meant that he has been exalted and placed in a more honorable position. In the same way we say that the bodies have changed in nature, that is to say that they have been exalted, that their essence has become purer; we see therefore that to sublimate is the same thing as to purify; this is what our Water does. This is how we must understand our philosophical sublimation on which many have been mistaken.

Now, our Water mortifies, illuminates, cleanses and vivifies; it first makes the black colors appear during the mortification of the body, then come many and varied colors, and finally whiteness. In the mixture of Water and the ferment of the body, that is to say body prepared, an infinity of colors appear. This is how our Magisterium is taken from one, is made with one, and it is made up of four and three are in one. Learn again, venerable Father, that the philosophers have multiplied the names of the Mixed Stone for the better.They said it is bodily and spiritual, and they did not lie, the Sages will understand. For it has a spirit and a body; the body is spiritual only in the solution and the spirit became corporeal through its union with the body. Some call it ferment, others Brass. Morien says: "The science of our Magisterium is comparable in everything to the procreation of man. First, coitus. Second, the design. Third, imbibition. Fourth, birth. Fifthly, nutrition or diet. I will explain these words to you. Our sperm which is Mercury, unites with the earth, that is to say with the imperfect body, called also Mother Earth (the earth being the mother of all the elements). This is what we mean by coitus. Then when the earth has retained in itself a little Mercury, it is said that there is has design. When we say that the male acts on the female, by which we must understand that Mercury acts on the earth. This is why the Philosophers said that our magisterium is male and female and that it results from the union of these two principles.

After the addition of Water, i.e. Mercury, the earth grows and increases while whitening, we then say there has imbibition. Then the ferment coagulates, that is to say, it joins the

imperfect body, prepared as has been said, until its color and appearance are uniform, this is birth, because 'at this moment appears our Stone whom the Philosophers have called: the King, as it is said in the Peat "Honor our King coming out of the fire, crowned with a diadem of gold; obey him until he has reached the age of perfection, feed him until he is great. his father is the Sun, its mother is the Moon; the Moon is the imperfect body.
The Sun is the perfect body.
Fifthly and lastly comes food, the more it is fed, the more it grows. However, lice feeds on her milk, that is to say, from the sperm which engendered it in the beginning. It is therefore necessary to imbibe it with Mercury, until it has drunk two parts of it, or more if necessary.

NOW PRACTICE
Follows.

Now let's move on to practice, as I announced above. And first all bodies must be brought back to the raw material to make transmutation possible. I will show you here all that was said above. So I beg you, O my son, do not disdain my Practice, because in it is hidden all our Magisterium, as I have seen it in my occult faith. Take a pound of gold, reduce it to very shiny filings, mix it with four parts of our purified water, grinding it and incorporating it with a little salt and vinegar, until combined. The gold having therefore been well amalgamated, put it in a large quantity of Eau-de-vle, that is to say of Mercury and put it all in the Urinai on our purified center; do below one very slow fire for a whole day; then let it cool, and when it is cold, take the Water and all that is with, filters through a linen cloth, until the liquid part has passed through the cloth. Put aside what will remain on the linen, collect it and having put in a fresh amount of Holy Water in the same vase as above, heats for a whole day, then filter as before. Repeat this until the whole body is converted into Water, that is to say into the raw material which is our Water.
This done, take all this Water, put it in a glass vase and cook over low heat until you see blackness appear on its surface; you will skillfully remove the black particles. Continue until the whole body is changed into a pure land. The more you repeat this operation, the better it will be. So anneal, removing the blackness, until darkness has disappeared, and Water, that is, our Mercury, appears bright. It is then that you will have the Earth and the Water
.

(H) Sweetly with great industry, separate the subtle from the gross

Then take all this earth, that is to say the darkness that you have collected; put her in a vessel glass, pour Holy Water over it, so that nothing rises above the surface of the water, let nothing float; and heated over low heat for ten days; then crushed and put back new Water; anneal the earth thus coagulated and thickened without adding water. Finally, cook over high heat, always in the same vase, until the clay becomes white and shiny. Having therefore whitened and coagulated our earth, take the Eau de vie which has been thickened with the help of a slight heat by the coagulated earth, cook it over a violent fire in a good cucurbit equipped with its capital, until all that there is a of water in the mixture has passed in the container and that the calcined earth remains in the gourd. Then take three parts for four of one ferment, that is, if you have taken a pound from the body imperfect or gold, you shall take three pounds of ferment,
i.e. Sun or Moon. You will first have to dissolve this ferment, reduce it to earth and repeat in a word the same operations as for the imperfect body. Only then will you unite them, you will soak them with the water that has passed through the container, and you will cook for three days or more. Soak in new, anneal and repeat this operation until these two bodies remain united, that is to say do more than one. You will weigh. Their color will not have not changed. Then you will pour on them the already named Water, little by little, until they absorb no more. In this union of bodies, the Spirit is incorporated into them and as they have been purified, it changes into their own nature. This is how the germ is transformed into bodies purified, which would not have happened before because of their coarseness and their impurities. The spirit grows in them, it increases and multiplies.

RÉCAPITULATION

Now, venerable Father, I will return to what I have said by applying it to the preparations of the ancient Philosophers and to their teachings, so obscure, so incomprehensible. However, weigh the words of the Philosophers, you will understand and you will admit that they have told the truth. The first word of our Magisterium or of the Work is the reduction of Mercury (the body), that is to say the reduction of copper or another metal to Mercury.
This is what the Philosophers call the solution, which is the foundation of the Art, as Franciscus says:
"If you do not dissolve the bodies, you labor in vain."

It is this solution of which Parmenides speaks. in the Peat of the Philosophers. On hearing the word of solution, the ignorant immediately think of the Water of the clouds. But if they had read our books, if they had understood them, they would know that our Water is permanent, and that separated from its body it therefore becomes immutable. So the solution of the Philosophers is not Water of the cloud, but it is the conversion of bodies into Water from which they were first procreated, i.e. in Mercury. In the same way the ice is changed into the water which had first given birth. Behold then that by the grace of God you know the first element which is Water and the reduction of this same body in the raw material.
The second word is "What is made of the earth". That's what the Philosophers said. "Water comes out of the earth. You will thus have the second element which is the earth. The third word of the Philosophers is the purification of the Stone. Morien says on this subject: "This Water putrefies and purifies itself with the earth, etc. " Philosopher says: "Join the dry with the wet; c, the dry is the earth, the wet is the Water." You will already have the Water and the earth in itself and the bleached earth with Water. The fourth saying is that Water can evaporate through sublimation or ascension. It becomes aerial again by separating from the earth with which it was previously coagulated and joined; and you will thus have the Earth, the Air and Water. This is what the Philosopher in the Peat says:

> "Blanch it and sublimate over high heat until it escapes a spirit which is Mercury. That is why it is called bird of Hermes and chicken of Hermogenes. You will find at the bottom a calcined earth, it is an igneous force, that is to say, of an igneous nature. You will therefore have the four elements, earth, fire and this calcined earth"

which is the powder of which Morien speaks"

> *"Do not despise the powder that is at the bottom because it is in a low place. It is the earth of the body, it is your sperm and in it is the crowning achievement of the Work."*

Then with the aforesaid earth put the ferment, this ferment that the Philosophers call the soul: and here is why: just as the body of man is nothing without his soul, likewise the dead earth or filthy body is nothing without ferment, that is, without its soul.
Because the ferment prepares the imperfect body, changes it in its own nature as has been said. There's no ferments other than the Sun and the Moon, these two neighboring planets approaching by their properties natural.
This is what makes Morien say:

> "If you don't do not wash, if you do not whiten the filthy body and you do not give it a soul, you will have done nothing for the Magisterium."

The spirit is then united with the soul and the body, it rejoices with them and settles down. Water alters, and what was thick, becomes subtle. Here is what Astanus says in the Peat of the Philosophers: "The spirit joins the bodies only when these have been perfectly purified of their impurities.
In this union appear the greatest miracles. for all imaginable colors show themselves then and the imperfect body takes, according to Barsen, the color of ferment, while the ferment itself remains unaltered. O Father full of piety, may God increase in you the spirit of intelligence so that you weigh well what I am going to say: the elements can only be generated by heir own sperm. Now this sperm is Mercury.
Consider man who can only be begotten with the help of sperm, plants which can only be born of a seed, as much as is needed for generation and growth.
There are some who, believing they are doing the best, sublimate the Mercury, fix it, unite it to other bodies, and yet they find nothing. Here is why: a sperm can change me, it remains as it was; and it produces its effect only when it is carried in the woman's womb. This is why the Philosopher Mechardus says; "If our Stone is not put into the womb of the female, to be nourished there, it will not increase. O my Father, here you are, according to your desire, in possession of the Stone of the Philosophers.

Glory to God.
Here ends Arnaud de Villeneuve's little treatise, given
' to the Pope, Benedict XI, in the year 1303.

(L) By this means, you shall have the glory of the whole world

RAIMONDI LULLII

CLAVICULA

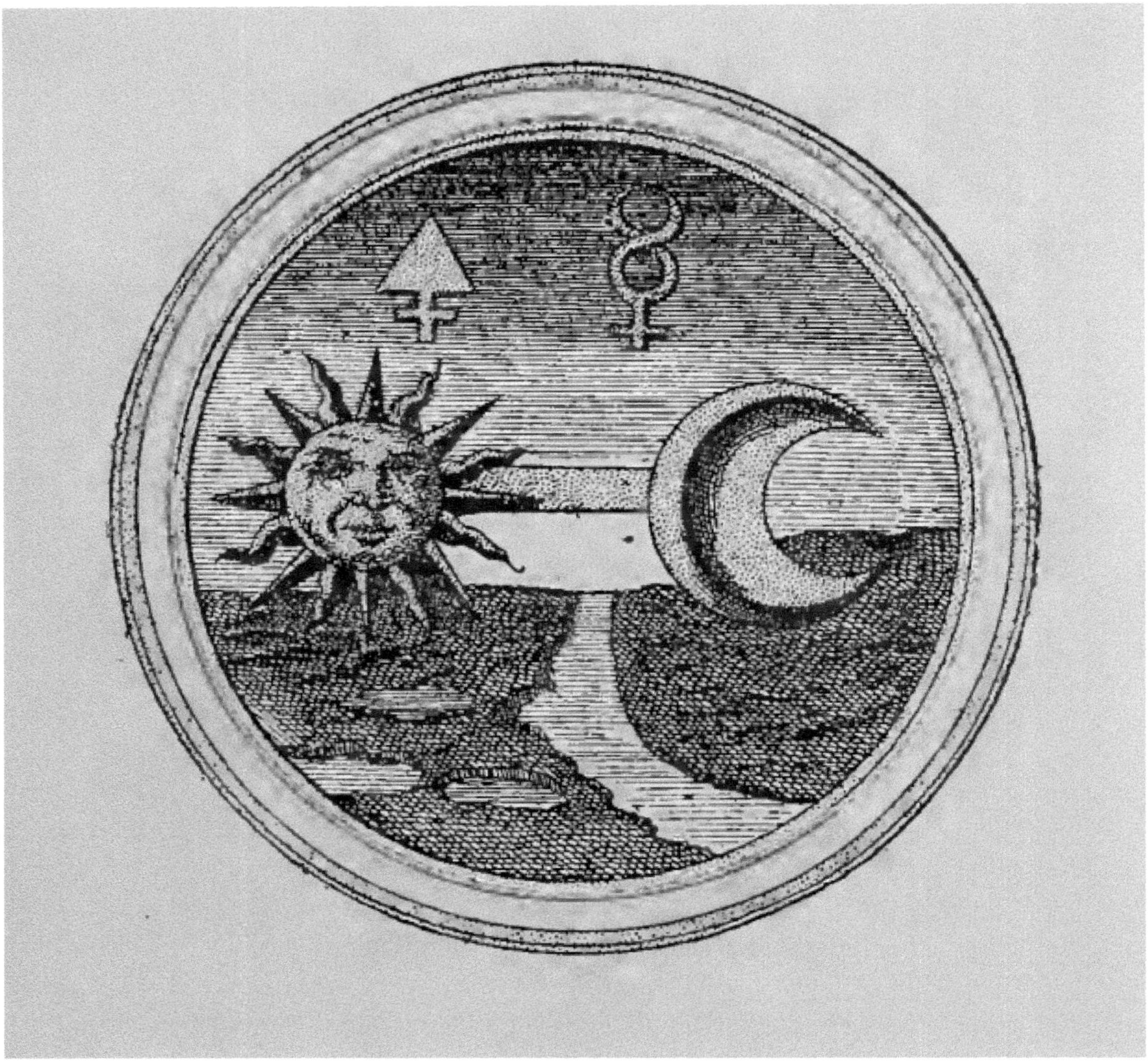

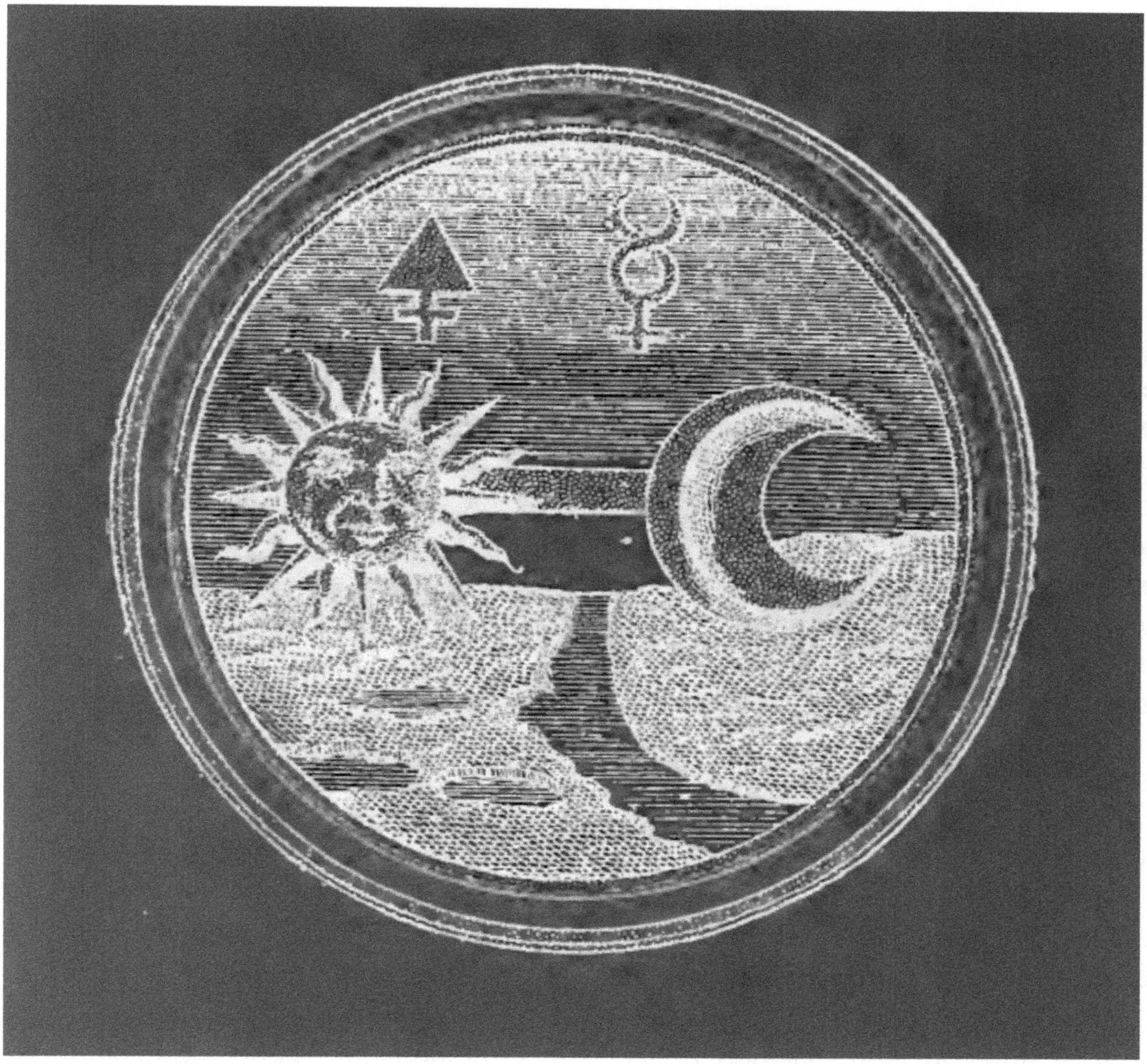

BIOGRAPHICAL NOTICE ON RAYMOND LULLE

Raymond Luile was born in Palma in the island of Majorca in 1235. His father, seneschal of Jacques !" of Aragon, destined him for a career in arms. The youth of R. Lulle was turbulent and licentious, the marriage did not alterhis conduct, but as a result of a violent love ended in an unfortunate way, he renounced the world and after having divided his property among his children, he retired to loneliness. It was then that he formed the project to convert the infidels, that will be the great idea to which he will devote his whole life. To learn Arabic, he buys a Muslim slave, but this one having guessed the purpose of his master, tries to assassinate him. Barely recovered, Raymond Luile founded a monastery where Arabic was taught, where missionaries are trained. Then he travels across Europe addressing popes, kings, emperors, asking some for their moral authority, others for help in silver to make his work fruitful. It was during these wanderings that he came into contact with Paris with Arnauld de'Villeneuve and Duns Scotus. He visits Spain, Italy, France, Austria. Joining the example to the word, he passes twice in Africa, is condemned to death in Tunis, and escapes only thanks to the protection of an Arab scholar who had taken a liking to him. In 1311, we find him at the Council of Vienna.

It was there that he received a letter from Edward II. This prince, showing himself favorable to his projects, R. Lully goes England. The king has him locked up in the Tower of London and forces him to do the great work. Raymond Lully changes into gold considerable masses of mercury and of tin, fifty thousand pounds," said Lenglet Dufresnoy. Of this gold we made the nobles with the rose or Raymondines Fearing for his life, R. Lulle escaped from London and returned to Africa. Barely landed, he begins to preach, the populace, indignant at his audacity, stoned him. The following night some Genoese picked him up, still breathing, from under a heap of stones and carried him aboard their ship, but he died within sight of Palma; he was buried in the convent of the Franciscans of this city (1315).

Main works: Codicillus seu vade mecum Testamenlum, Mercuriorum liber, Clavicula, Experimenta, Poleslas dmtiarum, Theoria et praclica, Lapidarium, Testaintum novissimum, etc. Clavicula seu Aperlorium is found in the Theatrum chimicum and in the Bibliolheca chemica Mangeii. As its name suggests, it is the key to all the other works of Raymond Lully.

(D) The earth is its nurse

RAYMOND LULLE'S CLAVICLE FROM MAJORCA

Treatise also known as the Universal Key, in which we will find clearly indicated all that is necessary to perfect the Great Work.

We called this work Clavicle, because without it, it is impossible to understand our others books, the whole of which embraces Art as a whole, because our words are obscure to the ignorant. I have made many treatises, very extensive, but divided and obscure, as can be seen from the Testament, where I speak of the principles of nature and all that has relates to art, but the text has been subjected to the hammer of Philosophy. The same applies to my book of the Mercure des philosophers, in the second chapter: on the fruitfulness of physical mines, the same for my book of the Quintessence of gold and silver, and finally for all my other works where art is treated in a way complete, except that I always hid the main secret. Now, without this secret no one can enter the mines of the philosophers and do something useful, that's why with the help and permission of the Most High to whom he has please reveal to me the Great Work, I will deal here with Art without any fiction. But beware of revealing this secret to the wicked; only share it with your friends intimate, although you should not reveal it to anyone, because it is a gift from God who makes it present to whom looks good to him. Whoever owns it will have a treasure eternal. Learn then to purify the perfect by the imperfect. The Sun is the father of all metals and the Moon is their mother, although the Moon receives its light from the Sun. Of these two planets depend on the whole magisterium. According to Avicenna, metals can only be transmuted after having been brought back to their raw material, which is true. You must therefore first reduce the metals to Mercury; but I do not mean here the vulgar, volatile Mercury, I speak of the fixed Mercury; because the vulgar mercury is volatile, full of a phlegmatic coldness, it is essential that it be reduced by the Mercury fixed, hotter, drier, endowed with qualities contrary to those of ordinary mercury. Therefore I advise you, my friends, to operate on the Sun and the Moon only after having led to their raw material which is sulfur and the Mercury of the philosophers. O my children, learn to use this material venerable, because I warn you under the faith of the oath, if you do not draw Mercury from these two metals, you will work like blind people, in the dark and in doubt. Therefore, O my sons, I conjure to walk towards the light, eyes open and not to fall blind into the abyss of perdition.

LA CLAVICULE, R.Llulli: CHAPTER I

DIFFERENCES OF VULGAR MERCURY AND MERCURY PHYSICAL.

We say: common mercury cannot be the Mercury of the Philosophers, by whatever artifice it has been prepared; for the vulgar Mercury can only hold out in the fire with the aid of a foreign corporeal Mercury which is hot, dry, and more digested than him. That's why I record our physical Mercury is of a warmer and more fixed nature than ordinary Mercury. Our Mercury corporeal becomes flowing mercury, not wetting the fingers; when it is joined to the vulgar mercury, they unite and join so well with the help of a bond of love, that it is impossible to separate them from each other, even from water mixed with water. Such is the law of nature. Our Mercury penetrates the vulgar mercury and mingles with it, drying up its phlegmatic dampness, taking away his coldness, which makes him black as coal and finally makes it crumble into dust.

Note that the vulgar Mercury cannot be used in place of our physical Mercury, which possesses natural heat to the desired degree; it is even for this reason that our Mercury communicates its own nature to ordinary Mercury. Much more, our Mercury, after its transmutation, changes the metals into pure metal, that is to say into the Sun and the Moon, as we have demonstrated in the second part of our Practice. But he does something more remarkable still, it changes the vulgar mercury into Medicine which can transmute imperfect metals into perfect ones. He changes vulgar mercury into true Sun and in real Moon, better than those coming out of the mine. Note again that our physical Mercury can transmute a hundred marcs and more, ad infinitum, all that that we will have, of ordinary mercury, unless it runs out. I also want you to know something else, Mercury doesn't mix easily and never perfectly.

To other bodies, if these have not previously been brought back to its natural species. This is why when you want to unite Mercury with the Sun or the Moon of the vulgar, you will first have to bring these metals back to their natural species which is ordinary mercury, using the bond of love. natural ; then the male unites with the female. Also our Mercury is active, hot and dry, while the vulgar Mercury is cold, humid, passive like the female who is kept at home in temperate heat until clouding. So these two mercurys turn black as coal; that's the secret real dissolution. Then they join together in such a way that it becomes impossible to ever separate them. They then appear in the form of a very white powder, and they beget male and female children by the true bond of love. These children will multiply to infinity according to their kind; because of an ounce of this powder, powder of projection, white or red elixir, you will make Suns in infinite number and you will transmute into Moon any kind of metal coming from a mine.

LA CLAVICULE, R.Llulli: CHAPTER II

EXTRACTION OF MERCURY FROM THE PERFECT BODY.

Take an ounce of cut Lunar-lime [fr. Lune coupellée], calcine it as described at the end of our work on the Magisterium. This lime will then be reduced to a fine powder on a porphyry plate. You will soak this powder, two, three, four times a day with good oil of tartar prepared in the manner described at the end of this work; then you will dry it in the sun. You will continue thus until the said lime has absorbed four or five parts of oil, the quantity of lime being taken as unity; you will pulverize the powder on the porphyry as it was said, after having dried it, because then it is more easily reduced to powder. When it has been well porphyrized, it will be introduced into a long-necked matrass. "You there add our menstrual stink made with two parts red vitriol and one part saltpetre; you will have previously distilled this menstrue seven times and you will have rectified it well by separating it from its earthy impurities, so that in the end this menstruation will be completely essential.

So we'll fight the matrass perfectly, we'll put it on over an ash fire, with a few coals, until you see the material boiling and dissolving. Finally we will distill on the ash until all the menstruation has passed and we will wait until the material is cold. When the vase is completely cold, we will open it, and the material will be placed in another very clean vase provided with its capital perfectly lute. We will place everything on ashes in a furnace. The lut being dry, we will first heat gently until all the water of the material on which one operates has passed through the recipient. Then increase the heat to dry out completely matter and exalt the stinking spirits that will pass into the capital and from there into the receptacle. When you see the operation arrived - at this point, you allow the ship to cool down, gradually reducing the fire. The vase being cold, you will draw from it the material that you will reduce to subtle powder on the porphyry. 'You will put the impalpable powder thus obtained in a well-fired and well-vitrified earthenware vase. Then you will pay by over plain boiling water, stirring with a clean stick, until the mixture is thick like some mustard. Stir well with the stick until you see a few globules of mercury appear in the matter. There's will soon have a big enough quantity according to what you have used of the perfect body, that is to say of the Moon. And until you have a large amount, pour from time to time boiling water and stir until all the material is reduced to a body similar to common mercury. We will remove the earthy impurities with cold water, we will dry on a cloth, we will pass through a skin of chamois. And then you will see wonderful things.

LA CLAVICULE, R.Llulli: CHAPTER III

OF THE MULTIPLICATION OF OUR MERCURY

In the name of the Lord. Amen.
Take three large Pure Luna in thin strips;

make an amalgam of it with four large well-washed common mercury.

When the amalgam is made you will put it in a small matrass having a neck of a foot and a half.

Then take our Mercury extracted above from the lunar body, and put it on the amalgam made with the perfect body and vulgar mercury;

fight the vase with the best possible lut and let dry.

Once this is done, shake the matrass vigorously to mix the amalgam well and Mercury.

Then place the vase where the material is in a small stove over a fire of a few coals only; the heat of the fire should not exceed to that of the sun when it is in the sign of the lion.

A stronger heat would destroy your material; also continue this degree of fire until the matter becomes black as coal and thick as mush.

Maintain the same temperature until where the material will take on a dark gray color;

when gray will appear, the fire will be increased by one degree and it will be twice as strong;

it will be maintained in this way until the material begins to whiten and becomes of dazzling whiteness.

We will increase the fire by one degree and we will maintain this third degree until the matter becomes whiter than snow and is reduced to a powder whiter and purer than ashes.

You will then have the Quicklime of the Philosophers and its sulphurous mine that the Philosophers have so well hidden.

(A) Its Father is the Sun

LA CLAVICULE, R.Llulli: CHAPTER IV

PROPERTY OF LIME OF PHILOSOPHERS.

This Lime converts an infinite quantity of vulgar mercury into a very white powder which can be reduced to real silver when united with some other body like the Moon.

LA CLAVICULE, R.Llulli: CHAPTER V

MULTIPLICATION OF THE LIME OF THE PHILOSOPHERS.

Take the vessel with the material, add to it two ounces of well-washed and dry common mercury; struggle with care, and put the vessel back where it was first. Regulate and govern the fire in degrees one, two, and three as above, until the whole is reduced to a very white powder; you can thus increase your Lime to infinity.

LA CLAVICULE, R.Llulli: CHAPTER VI

REDUCTION OF QUICKLIME IN REAL LUNA

Having therefore prepared a large quantity of our quicklime or mining lime, take a new crucible without its lid; put in it an ounce of pure Moon, and when it is melted, add four ounces of your powder agglomerated into pills. These little balls each weigh a quarter of an ounce. We throw them one by one on the Moon fusion, while continuing a violent fire until all the pills are melted; increase the heat even more so that everything mixes perfectly; finally we will flow in an ingot mold. You will thus have five ounces of fine silver, purer than natural; you will be able to multiply your physical mining at your thank you.

LA CLAVICULE, R.Llulli: CHAPTER VII

OF OUR GREAT WORK IN WHITE AND RED

Reduce to Mercury, as said above your Quicklime drawn from Luna.
This is our secret Mercury.

So take four ounces of our lime, extract Mercury from Luna as you did above.
You will collect at least three ounces of Mercury that you will put in a small long matrass collar as it has been said.

Then make an amalgam of one ounce of true Sun with three ounces of vulgar mercury and put it on the Mercury of the Moon.

Shake vigorously to mix well.

Fire the vessel carefully and put it in the furnace, adjusting the fire to the first, second and third degree.

In the first degree, matter will turn black like coal ; we then say that there is a solar eclipse and Moon. It is the true conjunction that produces a child, Sulphur, full of temperate blood.

After this first operation, we continue with the fire of the second degree until the matter is gray.

Then we go to the third degree until the moment when the material appears perfectly white.

We increase then fire until the matter turns red like cinnabar and is reduced to red ashes
.
You can reduce this Lime to a very pure Sun, doing the same operations as for the Moon.

LA CLAVICULE, R.Llulli: CHAPTER VIII

HOW TO CHANGE

THE ABOVE STONE INTO A MEDICINE THAT TRANSMUTES ANY KIND OF METAL INTO TRUE *SUN* AND TRUE *MOON* AND ESPECIALLY MERCURY-VULGAR IN METAL PURER THAN THE ONE THAT COMES OUT MINES.

After its first resolution our Stone multiplies one hundred parts of prepared matter, and after the second, a thousand.

We multiply by dissolving, coagulating, sublimating, fixing our matter which can thus increase indefinitely in quantity and quality.

So take from our white mineral, dissolve it in our stinky menses, which is called vinegar white in our Testament, in the chapter where we say:
"Take some good dry wine, put the Moon in it, that is to say Green Water and C, that is to say Saltpetre...." But let's not get lost;

take four ounces of our quicklime and dissolve in our menstruation, you will see it resolve into green water.

Else share in thirteen ounces of that same stinky menses you dissolve four ounces of well-washed common mercury, and as soon as the dissolution is complete, you will mix both solutions; put them in a well-sealed vessel, digest in horse manure for thirty days, then distill in a bain-marie until it passes nothing.

Redistill over a charcoal fire to extract the oil and then the material that will remain, will be black.

Take this one and distill for two hours on the ashes in a small stove.

The vase being cold, open it and pour into it the water that has been distilled above in a bain-marie.

Wash the material well with this water. Then distill the menses in a bain-marie; collect all the water,

add it to the oil and distil over the ashes, as has been said.

Repeat this operation until the matter remains at the bottom of the vessel, black as coal.

Son of science, then you will have the **Raven's Head**
that the Philosophers have so long sought, without which the Magisterium cannot exist.

Therefore, O my Son, remember the divine Supper of Our Lord Jesus Christ who died, was buried, and on the third day came back to light on the eternal earth. Know well, O my Son, that no being can live if he is not dead first.

So take your black body,
burn it in the same vessel for three days,
then leave chill.

Open it and you will find a spongy, dead earth,
which you will keep until needed.

to unite the body to the soul.
You will take the water that has been distilled in a bain-marie,
you will distill it several times in succession,
until it is well purified and reduced to a crystalline material.

Soak up your body which is the Black Earth with its own water, watering it little by little and heating it all, until the body becomes white and resplendent.

The vivifying water which clarifies has penetrated the body. The ship having been fought, you will heat up violently for twelve hours,- as if you wanted to sublimate the mercury vulgar.

The vase having cooled, you will open it and you will find your sublimated, white matter, it is our *Sigillated Earth*, it is our *sublimated body, raised to a*
high dignity, it is our Sulphur, our Mercury, our
Arsenic, with which you will warm our gold, is *our leaven, our quicklime* and it generates *in itself the Son of fire who is the Love of the philosophers.*

(N) Add to Liquor hepatis

LA CLAVICULE, R.Llulli: CHAPTER IX

MULTIPLICATION OF THE ABOVE SULFUR

Put this material in a strong matrass and pour over it an amalgam made with Quicklime from the first operation, the one we were reducing to silver. This amalgam is made with three parts of common mercury and part of our Lime; you will mix and you heat on the ashes. You will see the material stir, then increase the fire and in four hours the material will become sulphurous and very white. When she has been fixed, it will coagulate and fix Mercury; an ounce of matter will change one hundred ounces of Mercury into true Medicine; it will then operate on a thousand ounces, and so on. continued to infinity.

LA CLAVICULE, R.Llulli: CHAPTER X

FIXING MULTIPLE SULFUR.

We will take the multiplied sulfur, we will place it in a matrass and we will pour over it the oil that had been set aside during the separation of the elements.
Oil will be poured in until the Sulfur is soft. Then we will melt on the ashes, by heating to the second and third degree, until the whiteness inclusive. Then we will open the vessel and we will find a crystalline, white plate. To try it, put a fragment of it on a hot plate, and if it runs without producing smoke it is good. Then project one part of it on a thousand of mercury and it will be completely transmuted into silver. But if the medicine had been infusible and had not flowed, put it in a crucible and pour oil over it, drop by drop, until the medicine flows like wax, and then it will be perfect and will transmute a thousand parts of mercury and more ad infinitum.

LA CLAVICULE, R.Llulli: CHAPTER XI

REDUCTION OF WHITE MEDICINE INTO RED ELIXIR.

In the name of the Lord, take four ounces from the blade aforesaid and dissolve it in the Water of the Stone, which you have preserved. When the dissolution is complete, put ferment in a bain-marie for nine days. So take two parts by weight of our Red Lime and add them to the vessel, you will ferment again for nine days. Then you will distill in a bain-marie in a still, then on the ashes, adjusting the fire to the first degree until the moment when the matter turns black. This is our second dissolution and our second eclipse of the Sun with the Moon, this is the sign of the true dissolution and of the conjunction of the male with the female.

Increase the fire to the second degree, so that the matter becomes yellow. Then the fire will be raised to the fourth degree until the material melts like wax and is of a hyacinth color. It is then a noble material and a royal medicine that quickly cures all diseases; it transmutes all kinds of metal into pure gold better than the ore natural. Now let us give thanks to the glorious Savior who in the glory of heaven reigns one and three in eternity.

LA CLAVICULE, R.Llulli: CHAPTER XII

MAGISTERIUM SUMMARY

We have demonstrated that everything contained in this treaty is genuine, for we have seen with our own eyes, we operated ourselves, we touched with our own hands. Now we will without allegories and briefly summarize our Work.

So we take the Stone that we have said,
we sublimate it with the help of nature and art,
we reduce it to Mercury.
To this Mercury we add the white body which is of a similar nature,
and we cook until we have prepared the real mineral.

This mining will multiply at your will.

The material will again be reduced to Mercury,
which you will dissolve in our Menses until the Stone becomes volatile and separated from all its elements.

Finally we will perfectly purify the body and the soul.
A heat natural and temperate will then make it possible to achieve the conjunction of body and soul.

The stone will become mining;

we will continue the fire until the material becomes white,

we then call it Sulfur and Mercury of the Philosophers;

it is then that by violence from the fire, the fixed becomes volatile,
insofar as the volatile will have rid itself of its gross principles and will have sublimated itself whiter than snow.

We will throw what remains at the bottom of the ship, because it is good for nothing.

Then take our Sulfur which is the oil we have already spoken about and you multiply it in the still until it is reduced to a powder whiter than snow.

We will fix the powders multiplied by nature and by art, with water, until when tested by fire they flow smokeless like wax.

It is then necessary to add the water of the first solution;
everything having dissolved, we will put in something yellow which is gold,
we will unite and distill the whole spirit.

Finally, we will heat to the first, second, third and fourth degree
until the heat reveals the true hyacinth color, and the material fixes either fuse.

You will project this matter on a thousand parts of vulgar mercury and it will be transmuted into fine gold.

LA CLAVICULE, R.Llulli: CHAPTER XIII

CALCINATION OF LUNA FOR THE WORK.

Take one ounce of Fine Cuppellee Luna and three ounces of Mercury.

Amalgamate, first heating the silver in lamellae in a crucible and

then adding the mercury;

stir with a chopstick, while continuing to heat through. We will then put this amalgam in vinegar with salt;

we'll crush everything with a pestle in a wooden mortar, while washing and removing impurities. We'll stop when the mix is perfect.

Then we wash with warm and clear ordinary water,

then we will pass through a very clean cloth. What remains on the linen being the most essential part of the body,

one will mix it with three parts of salt, grinding well and washing.

Finally, we will calcine for twelve hours.

We will start grinding again with salt, and this three times,
each time renewing the salt.

Then we will pulverize the material so as to obtain an impalpable powder;

wash in hot water until all salty flavor is gone.

Finally we will pass through a cotton filter,

we will dry out, and we will have white Lime. We will put it in reserve, to be sure.

serve when needed, lest the humidity spoil it.

LA CLAVICULE, R.Llulli: CHAPTER XIV

PROCESS FOR PREPARING OIL FROM TARTAR.

Take good tartar, whose fracture is brilliant,

calcine it in the reverberatory furnace for ten hours;

then you will put it on a marble slab after have sprayed it and you will leave it in a humid place,

it will dissolve into an oily liquid.

When it is completely liquefied, it will be passed through a cotton filter.

You will keep it carefully, it will serve you to soak your lime.

LA CLAVICULE, R.Llulli: CHAPTER XV

STINKY MENSTRUE TO REDUCE OUR QUICKLIME IN MERCURY AFTER DISSOLVING IT WHEN IT HAS ALREADY BEEN SOAKED WITH SCALE OIL.

Take two pounds of vitriol, one pound of saltpetre, and three ounces of cinnabar.

We redden the vitriol,

we pulverize it,

then we add the saltpeter and the cinnabar,

we grind all these materials together, and put in a well-lit distilling apparatus.

We first distill on a slow fire, it is absolutely necessary.

This water will distill leaving its impurities which will remain at the bottom of the curcurbite and thus you will have this excellent menstrue.

LA CLAVICULE, R.Llulli: CHAPTER XVI

OTHER MENSTRUE TO SERVE AS A SOLVENT FOR THE ROCK.

Take three pounds of red Roman vitriol, one pound of saltpeter, three ounces of cinnabar,

grind all these materials together on the marble.

Then put them in a large and solid matrass,

add some brandy

ground seven times, then

seal the vessel perfectly and put it for fifteen days in manure of horse.

Then we will gently distill so that all the water goes into the container.

Then we will increase the fire until the capital is brought to white;

We then allow to cool.

We will remove the container

we will close it perfectly with wax and we will keep it.

Notice that this menstruation must be rectified seven times, each time rejecting the residue. Only then will he be good for the work.

(T) Transfer the resulting liquid to a clean container

Finis
LA CLAVICULE, R.Llulli
———------------------

ROGERII BACHONIS
SPECULUM ALCHEMIÆ

—

ROGER BACON
THE MIRROR OF ALCHEMY

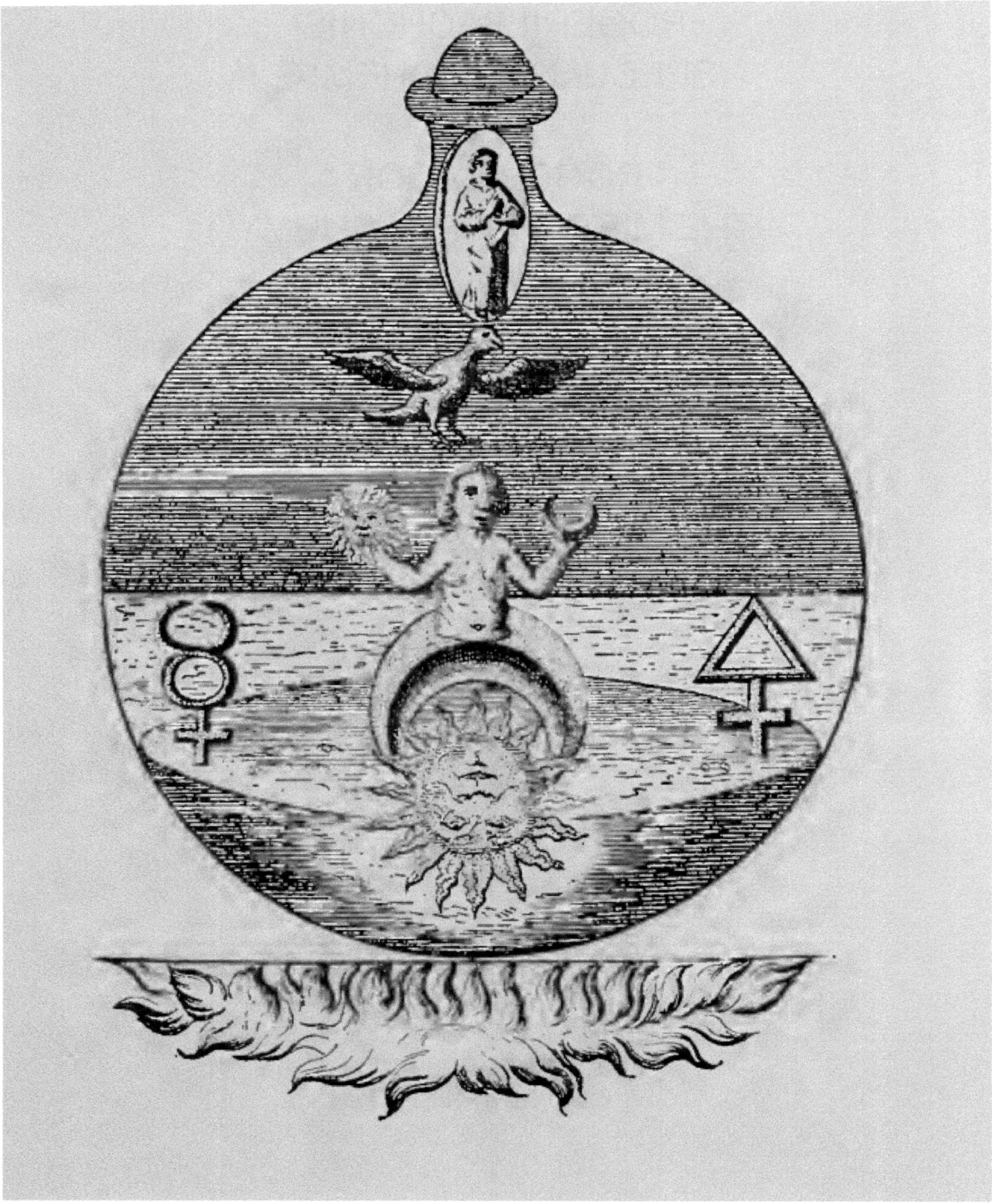

BIOGRAPHICAL NOTICE ON ROGER BACON

Roger Bacon was born in 1 2 14 in Ilcester, County of Sommerset. He did his first studies at Oxford, and then came to Paris to take the titles of Master of Arts and doctor of theology. At that time, Albert the Great professed publicly in Paris. Back in England, he entered the Franciscan Order around 1240.

He learned Greek, Arabic, Hebrew to read the ancients authors in the text. He thus acquired a prodigious erudition. He returned to Paris, which offered him more facilities for his studies. His ignorant superiors, afraid of his science, began to persecute him. Clement IV who admired him was powerless to protect him, and Bacon had to hiding from his superiors to write and send to the pope the Opus majus. Nicholas III succeeded Clement IV.

It was under this pontiff that Jérôme d'Esculo, general of the Franciscans, passing through Paris, had Roger Bacon locked up, accusing him of magic and heresy. Jerome d'Esculo was himself elected pope under the name of Nicolas IV, and Roger Bacon despaired of ever getting out of his dungeon when Raymond Gaufredi was named General of the Franciscans. A gentle and learned man, Raymond set Roger Bacon and several other Franciscans free. Bacon returned to England, but he had suffered too much, he was too old to resume his cherished studies. He died at Oxford 00-1294; on his deathbed he let fall these sad words: "I repent of having given so much trouble in the interest of science! » The works of R. Bacon relating to alchemy have been brought together in a collection entitled: Rogerii Baconis Thésaurus chimiçus, un vol. in-8°. Francofurti, 1603 and 1620.

List of treatises by Roger Bacon:

Alchimia major,
Brepiarium de dono Dei,
De leone piridi,
Secreiumsecreiorum,
Speculum alchemice,
Episiola de secrelis operibus arlis et natures ac nullitale magic.

This treatise ,that you read now, can be found in Latin in the Bibliotheca chemica mangeti, in the Thesaurus chimicus, in volume II of the Theatrum chimicum, it is from this text that the present translation was made. It is a treatise on speculative or theoretical alchemy.

A SMALL TREATISE OF ALCHEMY BY ROGER BACON ENTITLED

MIRROR OF ALCHEMY

PREFACE

In their writings the Philosophers expressed themselves from many different ways, but always enigmatic. They bequeathed to us a noble science among all, but veiled completely from us by their word cloudy, completely hidden under an impenetrable veil. And yet they were right to do so. Also, I conjure you to exercise your mind with perseverance on these seven chapters, which contain the art of transmuting metals, without having to worry about the writings of other philosophers. Iron often in your spirit their beginning, their middle, their end, and you there find inventions so subtle that your soul will be filled with joy.

(Q) Stir the mixture thoroughly

CHAPTER I
DEFINITIONS OF ALCHEMY

In some ancient manuscripts, we find several definitions of this art which it is important to discuss here. Hermes says; "Alchemy is the immutable science which works on bodies with the aid of theory and experience, and which, by a natural conjunction, transforms them into a more valuable superior species.

Another philosopher said: "Alchemy teaches to transmute any species of metal into another, that with the aid of a particular Medicine, as can be seen from the numerous writings of the Philosophers. This is why I say; "Alchemy is the science which teaches how to prepare a certain medicine or elixir, which being projected on imperfect metals, gives them perfection in the very moment of projection.

CHAPTER II
NATURAL PRINCIPLES AND THE GENERATION OF METALS

I will speak here of the natural principles and the generation of metals. First note that the principles of metals are Mercury and Sulphur. These two principles have given rise to all metals and all minerals, of which there are however a large number of different species. I say further that nature has always had as its aim and ceaselessly strives to arrive at perfection, at gold. But as a result of various accidents which hinder its progress, the metallic varieties are born, as is clearly stated in several philosophers. According to the purity or the impurity of the two component principles, that is to say Sulfur and Mercury, there are produced perfect or imperfect metals, gold, silver, tin, lead, copper, iron. Now piously collect these teachings on the nature of metals, on their purity or their impurity, their poverty or their richness in principle.

Nature of Gold:
> Gold is a perfect body composed of a pure, fixed, brilliant, red Mercury and of a Sulfur pure, fixed, red, noncombustible. Gold is perfect.

Nature of Silver:
> it is a pure, almost perfect body, composed of a pure, almost fixed, brilliant, white Mercury. Its Sulfur has the same qualities. There is no lack of
> Silver only a little more fixity, color and weight.
> Nature of the pewter; it is a pure, imperfect body,
> composed of a pure Mercury, fixed and volatile, brilliant,
> white on the outside, red on the inside. Its Sulfur has the same qualities. The tin only lacks to be a little more cooked and digested.

Nature of lead:
> it is an impure and imperfect body,
> composed of an impure, unstable, terrestrial, powdery Mercury, slightly white on the outside, red on the inside. His Smile is similar and more combustible.
> Lead lacks purity, fixity, color; he is not cooked enough.

Nature of copper:
> copper is an impure and imperfect metal, composed of an impure, unstable, terrestrial, combustible, red, lusterless Mercury. The same applies to
> its Sulphur. Copper lacks fixity, purity, weight. It contains too much impure color and incombustible earthy parts.

Nature of iron:
> iron is an impure, imperfect substance, composed of an impure Mercury, too fixed, containing combustible earthy parts, white and red, but without glow. It lacks fusibility, purity, weight; it contains too much impure fixed Sulfur and combustible earthy parts. Any alchemist must take into account the above.

CHAPTER III
FROM WHICH ONE MUST WITHDRAW THE NEXT MATTER OF THE ELIXIR

In what precedes we have sufficiently determined the genesis of perfect and imperfect metals. Now we will work to make imperfect matter pure and perfect. It appears from the chapters precedents that all metals are composed of Mercury and Sulphur, that the impurity and imperfection of components is found in the compound; as we can add to metals only substances drawn from themselves, it follows that no foreign matter can serve us, but that whatever is composed of two principles, suffices to perfect, and even to transmute the metals.

It is very surprising to see people, however skilful, to work on animals, which constitute a very remote matter, whereas they have at hand a sufficiently close matter in minerals. It is not impossible that a Philosopher placed the Work in these remote matters, but it is by allegory that he will have done it. Two principles make up all metals and nothing can attach itself, unite with metals or transform them, if it is itself composed of the two principles. It is as reasoning forces us to take for Matter of our Stone, Mercury and Sulphur. Mercury alone, Sulfur alone cannot engender metals, but by their union, they give birth to various metals and to numerous minerals. So it is obvious that our Stone must be born from these two principles.

Our last secret is very precious and very hidden: on what mineral matter, next among all, should we operate directly? We are forced to choose carefully. Suppose first that we derive our matter of plants: grasses, trees and all that
arises from the earth. It will be necessary to extract the Mercury and the Sulfur by a long cooking, operations that we let us push back, since nature offers us ready-made Mercury and Sulfur.

If we had elected animals, we would have to work on human blood, hair, urine, excrement, chicken eggs, finally all that can be obtained from animals.
There again, we would have to extract Mercury and Sulfur by cooking. We reject these operations for our first reason. If we had chosen mixed minerals, such as the various species of magnesia, marcasites, tuties, rosacea or vitriols, alums, borax, salts, etc., it would even be necessary extract Mercury and Sulfur from it by cooking, which we reject for the same reasons as above. If we choose one of the seven spirits as Mercury alone, or sulfur alone, or else Mercury and one of the two sulphurs, either quick-sulphur, or orpiment, or yellow arsenic, or red arsenic, we do not could perfect them, because nature does not only perfects the determined mixture of the two principles.

(S) Filter the mixture to separate any solids

We can't do better than nature, and we would have to extract from these bodies the Sulfur and the Mercury, what we repel as above.

Finally, if we took the two principles themselves, we would have to mix them in some unchangeable proportion, unknown to the human mind, and then cook them until they were coagulated into a mixture.A solid mass.

This is why we reject the idea of taking the two separate principles, that is to say Sulfur and Mercury, because we do not know their proportion and that we shall find bodies in which the two principles are united in right proportions, coagulated and conjoined according to the rules.

Hide this secret well:
    Gold is a perfect and male body without superfluity or poverty.
    If he perfected the imperfect metals melted with him, it would be the red elixir.
    Silver too is an almost perfect and female body, and if by simple fusion it made imperfect metals almost perfect, it would be the white elixir. What is not and what cannot be, because these bodies are perfect to a single degree. If their perfection were communicable to imperfect metals, the latter would not be perfected and they would be the perfect metals which would be defiled by contact with the imperfect. But if they were more than perfect, double, quadruple, a hundredfold, etc., they could then perfect the imperfect.

Nature always operates simply, that is why perfection is simple in them, indivisible and not transferable. They could not enter into the composition of the Stone, as ferments, to shorten the work; they would in fact be reduced to their elements, the sum of the volatile exceeding the sum of the fixed. And because gold is a perfect body composed of a brilliant red Mercury and a similar Sulphur, we shall not take it as the material of the Stone for the red elixir; because it is too simply perfect, without subtle perfection, it is overcooked and naturally digested and we can barely work it with our artificial fire; the same for silver.
When nature perfects something, it does not however does not know how to purify it, perfect it intimately, because it operates with simplicity. If we choose
gold and silver, we could hardly find a fire able to act on them. Although we know
this fire, we cannot however arrive at the perfect purification because of the power of their bonds and their natural harmony; also push back the gold for
the red elixir, the silver for the white elixir.
We would find a certain body, composed of Mercury and Sufficiently pure sulfur, on which nature will have worked little. We pride ourselves on perfecting such a body with

our artificial fire and knowledge of the art. We will subject it to suitable cooking, purifying it, coloring it and fixing it according to the rules of the art. Must therefore choose a matter which contains a pure, clear, white and red Mercury, not completely perfect, mixed equally, in the desired proportions and according to the rules, with a Sulfur similar to it. This material must be coagulated into a solid mass and such that with the aid of our science and our prudence, we can arrive at to purify it intimately, to perfect it with our fire, and to make it such that at the end of the Work, it is thousand times purer and more perfect than ordinary bodies cooked by natural heat.

(V) Add additional ingredients as necessary

So be careful; for if you have exercised subtlety and
sharpness of your mind on those chapters where I manifestly revealed to you the knowledge of Matter, you now possess this thing, ineffable and delectable,
object of all the desires of the Philosophers.

CHAPTER IV
OF THE MANNER OF REGULATING THE FIRE AND OF MAINTAINING IT.

If your head is not too hard, if your mind is not enveloped completely in the veil of ignorance and intelligence, I can believe that in the preceding chapters you have found the true Matter of the Philosophers, matter of the Blessed Stone of the Wise, on which Alchemy will operate in order to perfect imperfect bodies with more than perfect bodies. Nature only offering us perfect or imperfect bodies, it is necessary for us to make it indefinitely perfect by our work the Matter named above. If we do not know how to operate, what is the cause, if not that we do not observe how nature perfects metals every day? Do we not see that in the mines the coarse elements cook and thicken so well? by the constant heat existing in the mountains, that with time it is transformed into Mercury? That the same heat, the same cooking transforms the fatty parts of the earth in Sulfur? That this heat applied for a long time to these two principles, engenders according to their purity or their impurity, all the metals? Do we not see that nature produces and perfects all metals by cooking alone? O infinite madness, who therefore, I ask you, who therefore obliges you to want to do the same
thing with the help of bizarre and fantastic diets? This is why a Philosopher said: "Woe to you who want to surpass nature and make metals more than
perfected by a new diet, the fruit of your insane stubbornness. God gave nature immutable laws, that is to say, it must act by continuous cooking,

and you fools, you despise her or you don't know
restrict. He says likewise, "Fire and azoth must
suffice. And elsewhere: "Heat perfects everything." And elsewhere: "You have to bake, bake, anneal and not get away with it."
tired. And in different passages: "Let your fire be calm and gentle; may it continue thus every day,
always uniform, without weakening, otherwise a
great pity. — Be patient and persistent. —Crushed seven times. "Know that all our magisterium is
made of one thing, the Stone, in one way, by cooking
and in one vessel. — Crushed fire. — The Work is similar to the creation of man. In
childhood we feeds him light foods, then when his bones have
strengthened, the food becomes more fortifying; likewise
our magisterium is first subjected to a light fire with
which must always be acted upon during cooking. But
although we constantly speak of moderate fire, we
let us nevertheless imply that in the regime of
the Work must be increased little by little and by degrees until the firi.

CHAPTER V
OF THE VESSEL AND THE FURNACE.

We have just determined the way of operating, we are now going to speak of the vessel and the furnace, to say how and with what they must be made. When nature cooks the metals in the mines with the help of natural fire, she cannot
only be achieved by using a vessel suitable for cooking. We propose
to imitate nature in the regime of fire, so let's imitate her
also for the ship. Let's examine the place where
metals. We see first obviously in a mine, that under the mountain there is
has fire, produce sant' an equal heat, the nature of which is to rise without
cease. As it rises it dries up and coagulates the thick water
and coarse, contained in the bowels of the earth, and
transforms it into Mercury. The unctuous mineral parts of the earth are baked, collected in the veins
of the earth and flowing through the mountain, they engender Sulphur. As can be
observed in the veins of the mines, the Sulfur born from the unctuous parts of the earth
meets the Mercury. Then the coagulation of the metallic water takes place. The heat
continuing to act in the mountain, the various metals appear after a time
very long. One observes in the mines a temperature

constant, we can conclude that the mountain which contains mines is perfectly closed off from all sides by rocks; for if heat could escape, metals would never be born.

If therefore we want to imitate nature, it is absolutely necessary that we have a stove similar to a mine, not by its greatness, but by a disposition particular, such that the fire placed in the bottom does not find no way out to escape when it rises, so that the heat is reverberated on the vase, carefully closed, which contains the matter of the Stone. The vessel should be round, with a small neck. He must be of glass or earth as resistant as glass; close the orifice hermetically with a cover and bitumen. In the mines, the fire is not in immediate contact with the matter of Sulfur and Mercury; this is separated from it by the land of the mountain. Likewise the fire should not be applied to the naked vessel that contains Matter, but this vessel must be placed in another closed vessel with as much care as him, so that an equal heat acts on Matter, above, below, wherever he will be necessary. This is why Aristotle says in the Lights, that the Mercury must be cooked in a triple vessel of very hard glass, or, what is better, of earth possessing the hardness of glass.

(B) Its mother is the moon

CHAPTER VI
ACCIDENTAL AND ESSENTIAL COLORS THAT APPEAR DURING THE WORK.

Having elected the Matter of Stone, you also know the certain way to operate, you know with the help of which mode the various colors are made to appear by firing the stone. A Philosopher said "So many colors, so many names. For each new color appearing in the Work, the Alchemists invented a different name. So in the first operation of our Pierre, we gave the name of putrefaction, because our

Pierre is then -black". a When you have found the blackness, says another Philosopher, know that in this blackness hides the whiteness, and it must be extracted from it. After the putrefaction, the stone reddens and we have said on that; "Often the stone reddens, yellows and liquefies, then coagulates before true whiteness. She dissolves, putrefies, coagulates, mortifies, vivifies, blackens, whitens, adorns itself with red and white, everything this by itself.

(It can also turn green, because a philosopher said: "Cook

until a green child appears, it is the soul of the stone. Another said, "Know that it is the soul that dominates during the greenness.)

It also appears before the whiteness the colors of the peacock, a philosopher speaks of it in these terms:

"Know that all the colors that exist in the Universe or that one can imagine, appear before the whiteness, only then comes true whiteness.

The body will be cooked until it becomes shiny like the eyes fish (bubbley) and then the stone will coagulate at the circumference. "

"When you see the whiteness appearing on the surface in the vessel, says a wise man, be certain that under this whiteness hides the red; you have to extract it, cook until everything is red. "

There's finally has between red and white a certain ashy color, of which it was said:

"After whiteness, you cannot make more mistakes, because by increasing the fire, you will arrive at a grayish color. "

Do not despise the ashes, says a Philosopher, for with the help of God it will liquefy. Finally appears the King crowned with the diadem red, IF GOD permits.

CHAPTER VII
HOW TO MAKE PROJECTION ON METALS IMPERFECT.

As I had promised, I treated until the end our Great Work, blessed Magisterium, preparation of the white and red elixirs. Now let's talk about the way of making the projection, complement of the Work, eagerly awaited and desired. The Red Elixir, turns yellow to infinity and transforms all metals into pure gold. The White Elixir whitens infinitely and gives metals the perfect whiteness. But you should know that there a metals further than others from perfection and, conversely, there are some which are nearer. Although all the metals are equally brought to perfection by the Elixir, those which are next, become perfect more quickly, more completely, more intimately than the others. When we have found the closest metal, we will discard all the others. I have already said what are the near and distant metals and which is closer to perfection. If you are wise and intelligent enough, you will find it, in a preceding chapter, indicated without detour, determined with certainty. There is no doubt that he who has exercised his mind on this Mirror will find by his work the true Matter, and will know on which body it is advisable to make the projection of the Elixir to arrive at perfection.
Our precursors who found everything in this art by their philosophy alone, show us enough and without allegory, the straight path, when they say:

"Nature contains Nature, Nature rejoices in Nature, Nature dominates Nature and is transformed into the other Natures."

Like draws near to like, for likeness is a cause of attraction; there's has philosophers who have transmitted to us a remarkable secret on this subject. Know that nature spreads rapidly in one's own body, whereas one cannot unite it with a foreign body. Thus the soul quickly penetrates the body which belongs to it, but it is in vain that you would like to make it enter another body. The similarity is quite striking; the bodies in the Work, become spiritual and reciprocally the spirits become corporeal; the fixed body is therefore become spiritual. Gold, like the Elixir, red or white, has been taken beyond what his nature allowed, it is therefore not surprising that it is not miscible with molten metals, when we are content to project it there.

It would be impossible thus to transmute a thousand parts for a.
Also I will deliver to you a large and rare secret:
you have to mix a part of Elixir with a thousand of closest metal, enclose the whole in a vessel suitable for the operation, seal hermetically and put in the furnace to fix.
First heat slowly,
gradually increase the heat for three days
until perfect union.
It is the work of three days.

You can then begin again to project part of this product onto a thousand next metals,
and there will be transmutation.
All you need for this is a day, an hour, a moment.
Let us therefore praise our ever-admirable God in eternity.

Finis
THE MIRROR OF ALCHEMY, ROGER BACON
—

(B) Its mother is the moon

PARACELSI
THESAURUS THESAURORUM
ALCHEMISTORUM

PARACELSIS
THE TREASURE OF TREASURES
ALCHEMISTS

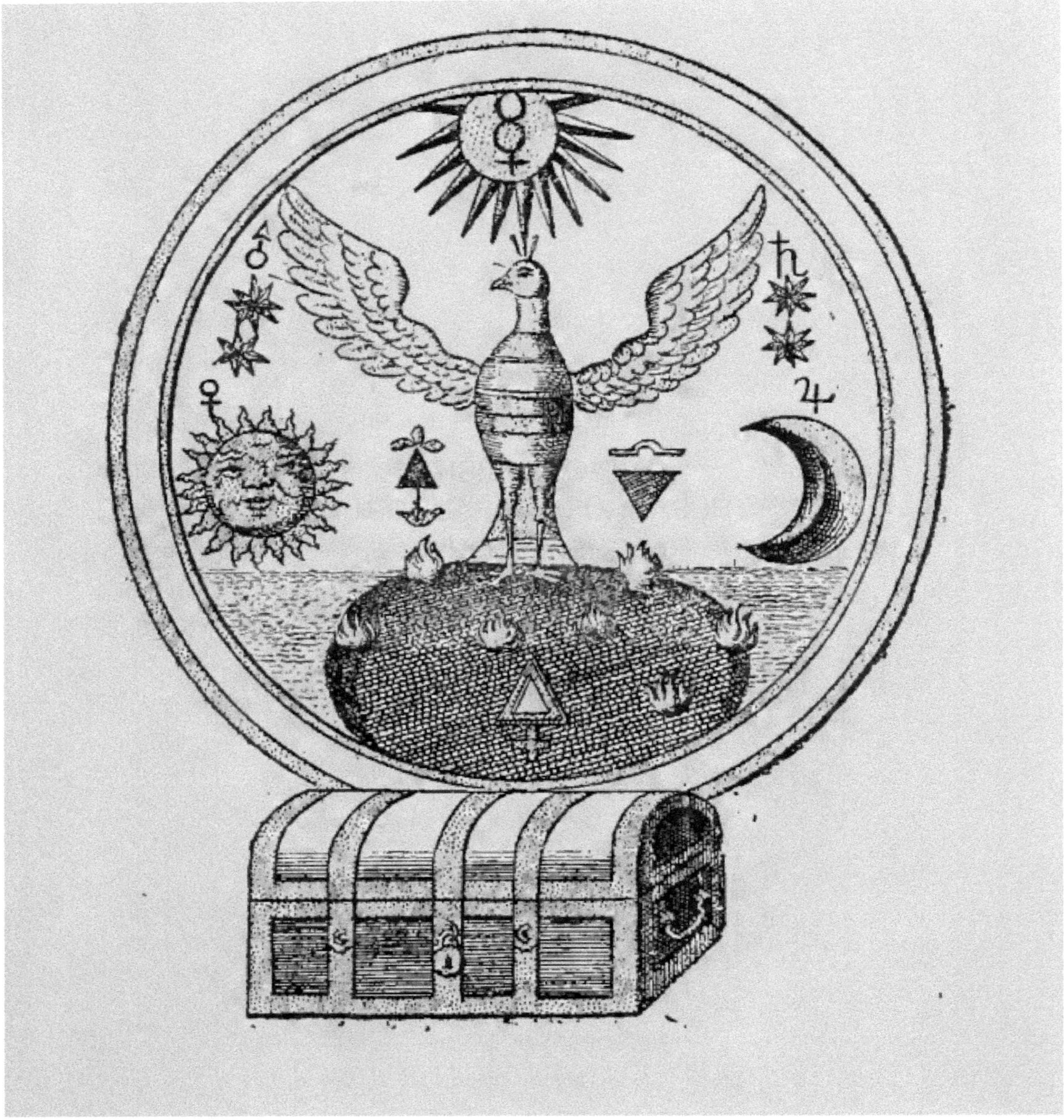

(U) Heat the liquid to a specific temperature

BIOGRAPHICAL NOTICE ON PARACELSE

Aureole Philippe Théophraste Paracejse Bombast ab Hohenheim, was born in 1493 at Einsiedeln, near Zurich, canton of Schwyz.
His father Guillaume, an educated doctor, taught him Latin, medicine and alchemy. The works of Isaac the Dutchman, which he read in his youth, gave him a deep love of alchemy. Since then he will never separate medicine from alchemy and it is the union of these two sciences which will characterize the school of paracelsists. His father sent him to finish his studies near Tritherne; this famous occultist had a great influence on the ideas of Paracelsus, because he taught him magic and astrology. Tritherne having retired to a convent, Paracelsus began to travel, he visited Portugal, Spain, Italy, France, the Netherlands, Saxony, Tyrol, Poland, Moravia, Transylvania, Hungary and Sweden. Perhaps he was even East, as he himself insinuates. He went through the cities and the villages, caring for the sick, selling remedies, drawing horoscopes, evoking spirits; on the other hand he interrogated old women, jugglers, gypsies, empirics, executioners, sorcerers. One communicated a secret to him, the other told him of a marvelous cure. Paracelsus collected everything, judging, comparing, observing. This is how he acquired his knowledge prodigious that the scholars of his time did not want to recognize, because it was found neither in Galen nor in Hippocrates.
In Hungary he entered the service of the Fuggers, bankers, alchemists and metallurgists; he could work as he pleased in their vast laboratories.
In i:;26, Oecolampadius Calls him to Basel to fill the chair of physics and surgery (of chemistry, says Haller). But he soon had to to leave the city, his violent teaching having attracted him enemies. He began to travel again, caring for princes and nobles, prelates and rich bourgeois. He died in 1541 in the Salzburg hospital.
Complete Works ; i" Paracelsi opera omnia medico, chemico, surgical, 3 vol. folio. Geneva, 1648: 2° Bûcher und Schrifleii Paracelsi, 10 vols. in 4°. Basel, 1589.
Treatises on alchemy:

Archidoxorum libri decem, —
De prœparaiionibus, —
De naiura rerum, —
De iinclura Physicorum, —
Cœlum Philosophorum, — Thesaurus
Ihesaurorum —
De mineralibus.
This treatise, translated for the first time into French, can be found on page 126, volume II of the Latin edition.

(M) Add Sulfuric Acid

THE TREASURE OF TREASURES OF THE ALCHEMISTS
BY PHILIPPE THEOPHRASTE BOMBAST, THE GREAT PARACELSO.

Nature engenders this mineral in the bosom of the earth.

There are two species that can be found in various localities in Europe.

The best I've had and who has been found good after assai is outside in the figure of the superior world, at the East of the solar sphere.

The second is found in the southern star and also in the first flower that the mistletoe of the earth produces on the star (i).

This passage is incomprehensible. So here is the latin text with ver batim translation:
(Optimum qitod mihi oblatum, ac in experimentado, genuine inventum est extra in figura majoris mundi, est in oriente astri sphœræ salis Alterum in Astro meridionali, jam in primo flora est, quem Visais terrce per suuni Astrum producit.)

[The most effective method that was suggested to me, and verified through experimentation, has proven to be authentic when applied in the broader context of the world. It lies in the eastern region, aligned with the prominent star within the realm of salt.]

After the first fixation it turns red;
in all the flowers and all the colors are hidden from it mineral.

The Philosophers wrote a lot about this, because it is of a cold and humid nature close to that of water.

For all things science and experience, the Philosophers who preceded me have targeted the Rock of the truth, but none of their strokes met the mark.

They believed that Mercury and Sulfur were the principles of all Metals, and they did not mention, even in a dream, the third principle.

However if by the spagyric art, one separates in addition to Water,
it seems to me that the Truth that I proclaim is sufficiently demonstrated;

neither Galen nor Avicenna knew it. If I had to describe for our excellent physicists the name, the composition, the dissolution, the coagulation, if I had to say how nature acts in beings since the beginning of the world, it would be enough for me barely a year to explain it and the skins of a whole herd of cows to write it down.

Now, I affirm that in this mineral, we find three principles, which are: Mercury, Sulfur and metallic Water which served to nourish it;

spagyric science can extract the metallic water from its own juice when it is not quite ripe, in the middle of autumn, even the pear on the tree. The tree contains the pear in power.

If the stars and nature agree, the tree first emits branches around March, then the buds grow, they open, the flower appears, and so on Until autumn when the pear ripens. It is the same for metals. They are born in a way similar in the bosom of the earth.

That the alchemists who seek the treasure of treasures note this carefully.
I will show them the way, the beginning, middle and end; in the following I will describe water, the sulfur and the particular balsam of the treasure.
By resolution and conjunction these three things will unite into one.

THE SULFUR OF CINNABAR.

Take some mineral cinnabar and operate like this. Cook it with rainwater in a stone vase for three hours ; then purify it carefully and dissolve it in aqua regia composed of equal parts of vitriol, nitre and sal ammoniac (another formula, vitriol, saltpetre, alum and common salt).

Distilled in a pot still while cohobing.

[Cohobation/ cohibation in alchemical laboratory work refers to a process of heating and sealing substances within a vessel to promote their intimate combination or union. It involves subjecting the materials to controlled heat while maintaining a closed system, allowing them to interact and undergo chemical transformations. The term "cohibation" is often used in the context of alchemical operations involving the union of different components or the dissolution of substances in a closed vessel. This technique aims to create an enclosed environment that encourages the desired reactions and facilitates the formation of new compounds. Cohobation is an important method employed in alchemical practices to achieve the desired transmutations and refine substances towards their perfected states.]

You will separate so carefully the pure from the impure. Then ferment for a month in horse manure. Then separate the elements according to the following: when the sign appears, start by distilling in the Alembic [still] with first degree fire.

Water and air will rise, fire and earth will remain in the background.

Cohobe and feed the still over the ash fire. Water and air will rise first,
then the element of fire, which skilled artists will easily recognize.

The Earth will remain in the bottom from the alembic you shall collect it;

many have sought it and few have found it.

You will prepare according to the Art ,this dead earth, in a reverberatory furnace,

then you will apply the fire of the first degree to it for fifteen days and fifteen nights.

This done you will apply the second degree to him for as many days and as many nights (your matter will have been enclosed in a hermetically sealed vessel).
You will finally find a volatile salt similar to a very light alkali, containing in itself the essence of fire and earth.

Mix this salt with the two elements you put aside, air and water.

Put heat on the ashes for eight days and eight nights, and you will find what many of artists have neglected.

Separate according to the rules of the spagyric art

> [Note:According to the spagyric art, the separation process involves the extraction and isolation of different components from a plant or substance. It consists of several steps:
>
> Maceration: The starting material is soaked or immersed in a suitable solvent, such as alcohol or water, to dissolve the desired constituents.
>
> Filtration: After maceration, the mixture is filtered to separate the liquid extract from the solid residue.

Distillation: The liquid extract is then subjected to distillation, a process that involves heating the solution to vaporize the volatile components. The vapors are collected and condensed to obtain a purified distillate.

Calcination: The solid residue left after filtration is subjected to high heat, or calcined, to remove impurities and obtain a purified ash.

Sublimation: Some substances undergo sublimation, a process where they are heated to convert directly from a solid to a vapor state and then condensed to obtain the purified substance.

Reconstitution: After the desired components have been separated and purified, they can be recombined in specific proportions to create a new, refined product. This process is known as reconstitution.

The spagyric art emphasizes the separation and purification of the different components of a plant or substance to obtain the purest and most potent essences. By isolating and recombining these components, practitioners aim to enhance their medicinal and energetic properties.]

and you will collect a white earth devoid of its dye.

Take the element of fire and the salt of the earth,

let the pelican digest to extract the essence.

It will separate once again into an earth that you will set aside.

(E) The father of all perfection in the whole world is here

THE RED LION

Then take the lion that passed first through the container as soon as you see its tincture, that is to say the fire that stands above water, air and earth. Separate it from its impurities by trituration. You will have then real potable gold. Sprinkle it with alcohol of wine to wash it; then distill in a still until you no longer taste the acidity of aqua regia. Then carefully enclose this sun oil in a hermetically sealed container. Heat to elevate it, so that it sublimates and splits.
Then place the vessel still tightly closed in a cool place. Reheat to rise, refrigerate to condense. Repeat this maneuver three time.
You will thus have the perfect tint of the sun. Save it for later.

THE GREEN LION

the air you had put aside.
Mix, putrefy for a month as it was said.

The putrefaction finished,
you will notice the sign of the elements.

Separate and you will soon see two colors, white and red.
Red is above white.

There red tincture of vitriol is so powerful that it
dyes all white bodies red, and all red bodies white, which is wonderful.

work on this tincture in a retort and you will see the blackness come out of it.

Put back into the retort what has been distilled, and start again until you obtain a white liquid.

Be patient and do not despair of the Work.

Rectify until you find the green lion, shining and true,
which you will recognize by its great weight.

It is the tincture of Gold.

You will see the signs admirable of our green lion, that none of the treasures of the Roman lion could not pay.

Glory to him who has known how to find it and draw the tincture from it.

It is the real balm natural of the celestial planets, it prevents putrefaction
bodies, and does not allow leprosy, gout, dropsy to implant itself in the human body.

When it has been fermented with the sulfur of gold, it is prescribed in the dose of a grain.

Ah! Charles the German, what have you done with your treasures of science? Where are your physicists? Where are your doctors? Where are these bandits who purge and medicate with impunity? Your firmament is turned upside down;

your stars, out of their orbits, wander far from the swampy path that had been laid out for them;

so your eyes were struck with blindness, as by an incandescent coal, when you beheld our splendor and our superb pride.

If your followers knew that their prince Galen (who is in hell) wrote me letters to recognize that I am right, they would make the sign of the cross with a fox's tail! And your Avicenna! he sits on the threshold of hell; I discussed with him his potable gold, physical tincture, mithridate and theriac. O hypocrites, who despise the truths taught you by a true doctor, instructed by nature, a son of God himself! Go always, impostors, who only prevail with the help of high protections. But be patient! after my death, my disciples will rise up against you, they will drag you in the face of the two, you and your dirty drugs, which serve you to poison the princes and the great ones of Christendom. Woe to your heads on the day of judgment! Me at contrary, I know that my kingdom will come. I will reign in honor and glory. It's not me who Praise is Nature, for she is my mother and I still obey her. She knows me and I know her. There light that is in it, I contemplated it, I demonstrated it in the Microcosm and I found it in the Universe. But I must return to my subject to satisfy the desires of my disciples, whom I willingly favor, when they are endowed with natural lights, when they know astrology and especially when they are skilful in philosophy, which teaches us to know matter of all.

Take four parts by weight of the metallic Water that I described,
two parts of the Earth of Red Sun,

One part of Sun Sulfur.
Put it all in a pelican, solidify and disintegrate three times.

You will thus have the Tincture of the alchemists. We will not speak here of its properties since they are indicated in the book of Transmutations.

With one ounce of Tincture of the Sun,
you can tint a thousand ounces of Sun;

if you possess the tincture of Mercury, you can likewise tint the body completely with vulgar mercury.

Likewise the tincture of "Venus will completely transmute into metal
perfected the body of "Venus.

All these things have been confirmed by experience. You have to hear the same
thing for the tinctures of the other planets: Saturn,
Jupiter, Mars, the Moon. Because from these metals we draw
also dyes; we will not say anything about it here, having
amply spoken of in the treatise on the Nature of Things and in the Archidoxes.
I have sufficiently described the spagyrlcs, the raw material of metals and minerals,
now they know the tincture of the alchemists.

It takes no less than nine months to prepare this tincture;
therefore work with ardor, without being discouraged;

During forty alchemical days, fix, extract, sublime, putrefy, coagulate in stone, and you will finally obtain the Phoenix of the philosophers.

But don't forget that the sulfur in cinnabar is an Eagle,
which flies without making wind,

and that it carries the body of the old Phoenix in a nest where it feeds on
the element of fire.

Her babies are tearing out her eyes, which produces whiteness.

This is the balm of his intestines which gives life to the heart, as have been taught the cabalists.

Finis
THE TREASURE OF TREASURES OF THE ALCHEMISTS By PARACELCUS

(O) Add Red Pulvis solaris

ALBERTI MAGNI
COMPOSITUM OF COMPOSITES

—

ALBERT THE GREAT
THE COMPOUND OF COMPOUNDS

7

CINQ TRAITÉS D'ALCHIMIE

7

BIOGRAPHICAL NOTE ON ALBERT GREAT

Albert the Great, of the ancient family of the Counts of Bollstadt, was born in Lavingen on the Danube in Swabia (1193). In his childhood, he was very unintelligent, but as a result of a vision his mind suddenly developed, and from that time he made rapid progress in all the branches of science. Around 1222, he entered the Order of Saint Dominic. He taught in the schools of the Order theology and philosophy. It was in Cologne that he distinguished Saint Thomas Aquinas among his students, they became close friends came together to Paris. Word of Albert the Great attracted such a crowd of listeners that he was obliged to teach in public squares; one of them has kept its name, it is the place Maubert or master Albert. In 1248 he returned to Cologne. For ten years he led in this city a peaceful existence conducive to study; provincial of his order, enjoying an uncontested authority with his contemporaries, helped by his monks in all the work he undertook, not having to worry about questions of silver how different his existence was from that of Roger Bacon! In 1259, Albert the Great was appointed bishop of Regensburg; but it was not long before he gave up the worries of the episcopate, and having resigned his office, he returned to the cloister. He died in Cologne in 1280 aged 87 years.

Complete Works:

Beaü Alberli, Ralisbonensis episcopi opera omnia, 21 vol. folio. Lugduni, 1651.

Alchemical treatises;
Libellas de Alchimia, —
Concordantia philosophorum de lapide philosophico, —
De rebusmetallicis, —
Composilum de composais, —
Breve compendium de orlu meiallorum.

This treaty, translated into English, can be found in the French translation in volume IV of the Thealrum chimicum, page 825. Hoeffer quotes in his History of Chemistry several passages of this treatise. Two of these passages are not found in the: Composilum de composais, but in the ; Libellas de Alchimia (Theat. chemistry., volume II).

Along with De Alchimia, this , that you are now reading, is the most important alchemical pamphlets by Albert the Great.

—

(F) Its force or power is entire if it be converted into earth

THE COMPOUND OF COMPOUNDS OF ALBERT GREAT

I will not hide a science revealed to me by the grace of God, I will not guard it jealously for me alone, lest I draw his curse. A science kept secret, a hidden treasure, what is their use?

The science that I learned without fictions, I transmit it to you without regrets.

Envy shakes everything, a man envious cannot be just before God. all science, all wisdom comes from God; it's a simple way to speak than to say that it comes from the Holy Spirit.

No cannot say: Our Lord Jesus Christ without implying; son of God the Father, by the operation of the Holy Spirit. Likewise, this science of truth cannot be separated from Him who communicated it to me.

I was not sent to everyone, but only to those who admire the Lord in his works and God deemed worthy. That he who has ears for hear this divine communication collect the secrets that have been transmitted to me by the grace of God and let him never reveal them to those who are unworthy.

Nature must serve as a basis and model for science, so Art works according to Nature as much that he can. The Artist must therefore observe Nature and operates as it operates.

(F) Its force or power is entire if it be converted into earth

CHAPTER I

ON THE FORMATION OF METALS IN GENERAL BY SULFUR AND MERCURY.

It has been observed that the nature of metals, such as we know it is to be engendered in a general way by Sulfur and Mercury. The only difference cooking and digestion produces variety in the species metallic. I myself observed that in one and the same vessel, that is to say in the same vein, nature had produced several metals and silver, scattered here and there. We have clearly demonstrated in our Treatise on Minerals that the generation of metals is circular, we pass easily from one to the other following a circle, neighboring metals have properties similar; that's why silver changes more easily into gold more than any other metal. There is indeed more to change in silver than the color and the weight, which is easy. For an already compact substance increases more easily by weight. And as it contains a yellowish-white sulphur, its color will also be easy to transform.

The same is true of other metals. Sulfur is, so to speak, their father and Mercury their mother.

This is even truer if we say that in the conjunction Sulfur represents the sperm of the father and that Mercury represents a coagulated menstruation to form the substance of the embryo. Sulfur alone cannot engender, thus the father alone. Just as the male begets from his own substance mixed with the menstrual blood, so the Sulfur begets with Mercury, but alone it produces nothing. By this comparison we want to imply that the Alchemist must first remove from the metal the specificity given to it by Nature, then proceed as nature proceeded, with Mercury and Sulfur prepared and purified always following the example of nature.

THE SULFUR.

Sulfur contains three moist principles.

The first of these principles is above all aerial and igneous, it is found in the outer parts of Sulphur, because of the great volatility of its elements, which fly away easily and consume the bodies with which they come into contact.

The second principle is phlegmatic, in other words watery, it is immediately placed under the previous one. The third is radical, fixed, adherent to internal parts.

This alone is general, and one cannot separate it from the others without destroying the whole building. The The fire is nothing else than the vapor of Sulphur; the vapor of well purified and sublimated Sulfur whitens and makes more compact. Also the skilful alchemists have custom to remove from the Sulfur its two superfluous principles by acid washes, such as the vinegar of the lemons, soured milk, goat's milk, urine of children. They purify it by leaching, digestion, sublimation. It must finally be rectified by resolution of so as to have nothing more than a pure substance containing the active, perfectible and proximate force of the metal. We here is in possession of a part of our Work.

(P) Add Natron

ON THE NATURE. MERCURY.

Mercury contains two superfluous substances,
land and water. The earthy substance has something
sulfur, the fire reddens it. The aqueous substance has a
unnecessary moisture.
Mercury is easily rid of its aqueous and earthy impurities by sublimations and
very acid washes. Nature separates it in a dry state from
SouAe and strips him of his land by the heat of the sun
and stars. She thus obtains a pure Mercury, completely stripped of its earthy substance, no longer containing
of foreign parties. She then unites it with a pure Sulfur
and finally produces in the bosom of the earth pure metals
and perfect. If the two principles are impure the metals
are imperfect. This is why in the mines we
finds different metals, which is due to the purification and variable digestion of their Principles. That
depends on cooking.

OF ARSENIC

Arsenic is of the same nature as Sulphur, both dye red and white.
But there see you moisture in arsenic,
and on fire it sublimates slower than sulfur.
We know how quickly sulfur sublimates and how it consumes all bodies except gold.
Arsenic can unite its dry principle with that of sulphur, they are tempered one another, and once united it is difficult to separate them; their tincture is softened by this union.
"Arsenic", says Geber, contains a lot of mercury, so it can be prepared like it.

Know that the spirit, hidden in sulphur, arsenic and animal oil, is called by philosophers White Elixir.

It is unique, miscible with the igneous substance, from which we derive the red Elixir; it unites with molten metals, as we have experienced, he purifies them, not only because of the aforementioned properties, but still because there is a common proportion between its elements.
Metals differ among themselves according to the purity or the impurity of the raw material, i.e. Sulfur and Mercury, and also according to the degree of the fire which engendered them.
According to the philosopher, the elixir is still called Medicine, because we assimilate the body of metals to the body of animals.
So we say that there is a hidden mind in sulphur, arsenic and oil extracted from animal substances.
This is the spirit we seek, with whose help we will dye all the imperfect bodies in perfect.
This spirit is called Water and Mercury by the Philosophers.

Mercury, says Geber, is a medicine composed of dry and wet, wet and dry. You understand the sequence of operations: extract earth from fire, air from earth, water from air, since water can resist fire.

Note these lessons are universal mysteries.
None of the principles that enter into the Work has power by itself;
for they are chained in the metals, they cannot perfect, they are no longer fixed.
They lack two substances, one miscible with molten metals, the other fixed which can coagulate and to stare.
Also Rhases said: "There has four substances that change over time; each of them is composed of the four elements and takes the name of the dominant element.

(C) In its belly, the wind hath carried it

Their wondrous essence has settled into one body and with the latter one can nourish the other bodies.

This essence is composed of water and air, combined in such a way that heat liquefies them. This is a marvelous secret. The minerals employed in Alchemy must, in order to serve us, have an action on the molten bodies. The stones we use are four in number, two dye white, the other two red. Also white, red, Sulphur, Arsenic, Saturn have only one same body.

But in this one body, how many obscure things!

And first of all it has no action on perfect metals.

In imperfect bodies, there is an acid, bitter, sour water, necessary for our art. Because it dissolves and mortifies the bodies, then revives them and recomposes them.

Rhases says in his third letter: "Those who seek our Entelechy, ask where the watery bitterness comes from."

Getting elementary.

We will answer them:

Impurity metals. Because the water contained in gold and silver is soft, it does not dissolve, on the contrary it coagulates and fortifies, because it contains neither acidity nor impurity like imperfect bodies.

This is why Geber has says:

"We calcine and dissolve useless gold and silver, because our Vinegar is drawn from four imperfect bodies; it is this mortifying and dissolving spirit which mixes the tinctures of all the bodies that we use in the work.

We only need this water, little

We care about other minds. "

Geber is right; we have nothing to do with a tincture that fire alters, on the contrary, the fire must give it excellence and strength so that it can combine with molten metals. It must strengthen, fix, and despite fusion remain intimately united to the metal. I would add that of the four imperfect bodies one can shoot everything. As to how to prepare Sulphur, arsenic and mercury, indicated above, it can be reported here. Indeed, when in this preparation we heat the spirit of sulfur and arsenic with acid waters or oil, to extract the igneous essence, the oil, smoothness, we take away what has superfluity in them; we are left with the igneous force and the oil, the only things that are useful to us; but they are mixed to the acid water which we used to purify, there is no way of separating them; but at least we are get rid of the unnecessary. We must therefore find another way to extract from these bodies, the water, the oil and the spirit, very subtle sulfur which is the real very active tincture that we seek to obtain.

So we will work these bodies by separating by decomposition or by
distillation of their natural component parts, and we we shall thus arrive at the simple
parts. Some, unaware of the composition of the Magisterium, want to work on
the only Mercury, claiming that it has a body, a soul, a spirit, and that it is the raw
material of gold and the silver. We must answer them that in truth a few
philosophers affirm that the Work is made of three things, mind, body and soul, drawn
from one. But on the other hand one cannot find in a thing what is not there. Now,
Mercury does not have the red tincture, therefore it does not
can alone suffice to form the body of the Sun; it would be impossible with Mercury alone
to bring the work to a successful conclusion. The Moon alone cannot suffice, however
this body is, so to speak, the basis of the work.
However one works and transforms the Mercury, it will never be able to constitute the
body. They also say: "There is found in Mercury a sulfur
red, so it contains the red dye. " Error !
Sulfur is the father of metals, we never find any
in mercury which is female. A passive matter cannot fertilize itself.
Mercury does contain Sulphur, but, like us
we have already said it is an earthly sulphur. notice
finally that Sulfur cannot support fusion; SO
the Elixir cannot be drawn from a single thing.

(J) Again, it descends to the earth

CHAPTER II

OF PUTREFACTION

Fire begets death and life. A light fire dries up the body. Here is the reason: the fire arriving at the
contact with a body, sets in motion the element similar to itself which exists in this body. This element is natural heat. This excites
the fire first extracted from the body; there's
a conjunction and the radical humidity of the body rises to its surface as long as the fire acts outside. As soon as radical humidity
which united the various portions of the body is gone,
the body dies, dissolves, resolves itself; all its parts separate from each other. The fire acts here as
a sharp instrument. Although it dries up and shrinks by itself, it can only do as much as there is.
has in the body a certain predisposition, especially if the body is compact as an element is. The latter lacks
of an agglutinating mixture, which would separate from the body after
Corruption. All this can be done by the Sun, because
that it is of a hot and humid nature compared to the
other bodies.

CHAPTER III

THE REGIME OF THE STONE. There's has four regimes of the Stone:
1° To decompose;
2° wash;
3° reduce;
4° fix.

In the first regime the natures are separated, for without division, without purification, there cannot be conjunction.

During the second regime, the separated elements are washed, purified, and brought back to the simple state.

On the third we change our Sulfur into a mine of the Sun, the Moon and other metals.

At fourth all bodies previously extracted from our Pierre, are united, recomposed and fixed to henceforth remain conjoined.

There are some who have five degrees in the Magisterium:
I° resolve substances into their raw material;
2° to bring our soil, that is to say the black magnesia to be close in nature to Sulfur and Mercury;
3° to make the Sulfur as close as possible to the mineral matter of the Sun and the Moon;
4° compose of several things an Elixir. white ;
5° perfectly burn the white elixir, give it the color of cinnabar, and from there, to make the Red Elixir.

Finally there are some which count four degrees in Work, other three, other two only. These last count thus:
1° implementation and purification of the elements;
2° conjunction. Note carefully what follows: the material of the Stone of Philosophers, is cheap; it is found everywhere, it is a viscous water like mercury that is extracted from Earth.

Our viscous water is everywhere, even in the Latrines, have said certain philosophers, and some fools, taking their words literally, have sought it in the excrement.

Nature operates on this matter by removing something from it, its earthy principle, and adding to it something, the Sulfur of the Philosophers, which is not the Sulfur of the vulgar, but an invisible Sulphur, tincture of red.

To tell the truth, it is the spirit of Roman vitriol.

Prepare it thus:
Take saltpeter and Roman vitriol,
2 pounds each; subtly crushed.

Aristotle is therefore right when he says in his fourth book of meteors. "All alchemists know that one cannot in any way change the form of metals, if one does not first reduce them to their raw material."

Which is easy as we will soon see. The Philosopher says that one cannot go from one extremity to the other without passing through the middle.

At one end of our philosopher's stone are two luminaries, gold and silver, at the other end the perfect elixir or tincture.

In the middle, philosophical eau-de-vie [water-of-life], naturally purified, cooked and digested.

All these things are close to perfection and preferable to bodies of nature more distant.

Just as, by means of heat, ice resolves into water, since it was once water, so metals resolve into their first matter, which is our water of life.

The preparation is indicated in the following chapters.
It alone can reduce all metallic bodies to their raw material.

(I) From the earth, it ascends to the heaven

CHAPTER IV

OF THE SUBLIMATION OF MERCURY.

In the name of the Lord, get yourself a pound of pure mercury from the mine.

On the other hand, take Roman vitriol and calcined common salt, ground and thoroughly mixed.

Put these last two materials in a large glazed earthenware vase over a low heat, until matter begins to melt and flow,

So take your mineral mercury, put it in a long-necked vase and pours drop by drop on the vitriol and molten salt.

Stir with a wooden spatula, until the mercury is completely devoured and there is no trace left.

When it has completely disappeared, dry the material over low heat overnight.

The next morning, you will take the well dried matter, you will grind it finely on a stone.

You will put the material pulverized in the sublimatory vase named aludel to sublimate it according to art.

You will put the capital (the top of the distillation unit) and you will coat the joints with philosophical lut or clay, so that the mercury cannot escape.

You place the aludel on its furnace and lute [fix it in place] it there so that it cannot bow and so that it stands straight

then you will make a small fire during four hours to drive out the humidity of mercury and vitriol;

after evaporation of moisture, you will increase the fire so that the white and pure matter of mercury separates from its impurities, this for four hours;

you will see if that is enough by introducing a wooden rod into the sublimatory vase through the upper opening, you will go down to the material and you will feel if the material, white mercury, is superimposed on the mixture.

If that is, remove the stick, close the opening of the capital with a lut so that the mercury cannot escape, and increase the fire so that the white matter of the mercury rises above the faeces, even in the aludel, for four hours.

Finally heat with wood in such a way as to obtain flames, the bottom of the vase and the residue must turn red;

keep on going thus as long as there remains a little white substance of mercury adhering to the faeces.

The force and violence of the fire will eventually separate it. So cease the fire, let cool the furnace and the material overnight.

The next morning remove the vase from the stove,
remove the read carefully so as not to soil the Mercury,
open the device;

if you find a white matter, sublimated, pure, compact, heavy, you have succeeded.

But if your sublimation was spongy, light, porous, pick it up, start the sublimation again on the residue by adding pulverized common salt again;

operate in the same vessel on his stove, in the same way, with the same degree of fire than higher.

Then open the vase, see if the sublimate is white, compact, dense, collect it and put it carefully aside to use it when you need it to complete the Work.

But if it did not yet present itself as it should be, you would have to sublimate it a third time until you obtain it pure, compact, white, heavy.

Note that by this operation you have removed two impurities from Mercury. First you took away all its superfluous moisture;
secondly you got rid of it of its impure earthy parts which remained in faeces; you have thus sublimated it into a clear, semi-fixed substance. Set it aside as recommended.

CHAPTER V

FROM THE PREPARATION OF THE WATERS FROM WHICH YOU SHALL DRAW WATER-LIFE.

Take two pounds of Roman vitriol,
two pounds of saltpetre,
one pound of calcined alum.
Crush well,
Mix perfectly,
put in a glass still, distill the water according to the ordinary rules, closing the joints tightly, lest the spirits escape.

Starts with a low heat, then heats up more strongly; then heat with wood until the device becomes white, so that all spirits distill.

So cease the fire, let the stove cool;
put this water carefully aside, for it is the solvent of the Moon;

keep it for the Work, it dissolves the silver and separates it from the gold.

She calcines Mercury and the crocus of Mars; it communicates to the skin of the man a brown coloring which goes away with difficulty.

It is the prime water of the philosophers, it is perfect in the first degree. You will prepare three books of this water.

SECOND WATER PREPARED BY SAL AMMONIAC

In the name of the Lord,
 take a pound of prime water and dissolved four lots of pure, colorless sal ammoniac;

once dissolved, the water changed color and acquired other properties.

Prime water was greenish, it dissolved the Moon, had no action on the Sun; but as soon as sal ammoniac is added to it, it takes on a yellow color, it dissolves gold, mercury, sulfur sublimated and imparts a strong yellow color to human skin. Keep this water carefully, because it will serve us in what follows.

THIRD-PARTY WATER PREPARED BY MEANS OF SUBLIMED MERCURY

Take a second pound of water and eleven lots of Mercury sublimated
(by Roman vitriol and salt) well prepared and quite pure.

Little by little you will pour the Mercury into the second water.

Then you will seal the mouth of the vial, fear that the spirit of Mercury will escape.

You will place the vial on lukewarm ashes, the water will immediately begin to act on the Mercury, dissolving it and incorporating it.

You will leave the vial on the hot ashes, there must not remain an excess of water and the sublimated Mercury must dissolve entirely.

Water acts by imbibition on Mercury until it has dissolved it.

If the water could not dissolve all the mercury,
you will take what remains at the bottom of the vial,
you will dry it on fire slow,
you will pulverize and dissolve it in a new quantity of second water.
You will repeat this operation until all the sublimated mercury has dissolved in the water.

You will unite in one all these solutions, in a very clean glass vase, the orifice of which you will close perfectly with oilcloth.

Set this carefully aside. For this is our third water, philosophical, thick, perfect in the third degree.
It is the mother of the Eau-de-vie [water-or-life] which reduces all bodies to their raw material.

(G) Separate the earth from the fire

QUARTER WATER WHICH REDUCES CALCINED BODIES INTO THEIR RAW MATERIAL.

Take tertian mercuric water, perfect in the third degree, limpid, and put it in the horse's belly to putrefy in a long-necked flask, clean, tightly closed,
for fourteen days.

Leave to ferment, the impurities fall to the bottom and the water changes from yellow to red.

At this time you will withdraw the vial and you will put it on ashes over a very low fire, adapt to it a capital of an alembic with its container.

Start the distillation slowly.

What passes drop by drop is our very limpid, pure, heavy eau-de-vie.
our Virginal Milk, our Very Sour Vinegar.

Continue the fire slowly until all the eau-de-vie [water-of-life] quietly distilled; then stop the fire, let the stove cool down and carefully store your distilled water.

This is our Eau-de-vie, Vinegar of the philosophers, Virginal milk which reduces bodies to their raw material.

It has been given an infinity of names.
Here are the properties of this water:
> a drop placed on a hot copper blade penetrates it immediately and leaves a white stain.
> Thrown on coals, it emits smoke;
> in the air it freezes and looks like ice cream.
> When this water is distilled, the drops do not all pass along the same path, but some pass here, others there.
> It does not act on metals like strong, corrosive water, which dissolves them, but it reduces to Mercury all the bodies it bathes, as you will see later.
> After putrefaction, distillation, clarification, it is pure and more perfect, freed from any igneous and corrosive sulphurous principle.
> It is not water that gnaws, it does not dissolve bodies, it reduces them to Mercury.
> It owes this property to Mercury primitively dissolved and putrefied in the third degree of perfection.
> It no longer contains any faeces or earthy impurities.

The last distillation separated them, the black impurities remained at the bottom of the still.

The color of this water is blue, limpid, red; put it aside. For it reduces all charred bodies and rotten in their radical or mercurial raw material.

When you want with this water to reduce the bodies calcined thus prepares the bodies.
Take a mark of the body you want, Sun or Moon; file it gently.
Pulverize these filings well on a stone with prepared common salt.
Separate the salt by dissolving it in hot water;
the pulverized lime will fall to the bottom of the liquid;
decant.
Dry the lime,
soak it three times with oil of tartar, each time letting the lime absorb all the oil;
put then the lime in a small flask;
pour over the oil of tartar, so that the liquid is two fingers thick,
then close the vial, put it in the horse's belly to putrefy for eight days;
then take the vial, decant the oil and dry the lime.
This done, put the lime in an equal weight of our brandy; close the flask and let it digest over a very low heat until all the lime has been converted into Mercury.
Then decant the water carefully, collect the corporeal Mercury, put it in a glass vase;
purify it with water and common salt,
dry it according to the menstruation,
put a fine cloth on it and express it in droplets. If it passes completely, that's good.
If there is anything left portion of the amalgamated body, coming from the fact that the dissolution was not complete,
put this residue with a new amount of holy water.
Know that the distillation of water must be done in a bain-marie;
for air and the fire, we will distill on the hot ashes.
The water must be drawn from the moist substance and not from elsewhere;
air and fire must be extracted from the dry substance and not from another.

PROPERTIES OF MERCURY

Properties of this Mercury.
It is less mobile, it runs less quickly than another mercury; he leaves traces of his body fixed in the fire; a drop placed on a red-hot slide
leaves a residue.

MULTIPLICATION OF THE PHILOSOPHICAL MERCURY

When you have your Philosophical Mercury,
take two parts and one part of the filings mentioned above;
make an amalgam by grinding everything together until perfect union.
Put this amalgam in a flask, close the orifice well and place on the ashes over a moderate fire.
Everything will be resolved in Mercury. You can thus increase it to infinity, because the amount of volatile exceeding always the sum of fixed, increases it indefinitely in communicating to it his own nature and there will always have enough.

Now you know how to prepare eau-de-vie, you know its degrees and properties, you know the putrefaction of metallic bodies, their reduction to matter first, the multiplication of matter ad infinitum.

I explained to you clearly what all the philosophers have hidden with care.

Practice of the Mercury of the sages.

It is not the mercury of the vulgar, it is matter first of the philosophers.
It is an aqueous, cold, humid element, it is a permanent water, it is the spirit of the body, greasy steam. Holy water. Etching. Water of the wise. Philosopher's Vinegar, Mineral Water. Dew of heavenly grace; it has many other names, and many that they are different, they all designate one and the same thing which is the Mercury of the philosophers;

It is the strength of alchemy; alone it can be used to make white and red dye, etc.
Take therefore in the name of Jesus Christ, our Most venerable,

Water of the philosophers, primitive Hyle of the sages; it is the stone that you were discovered in this treatise, it is the raw material of the perfect body, as you have guessed.

Put your material in a furnace, in a clean, clear, transparent, round vessel, whose orifice you will hermetically seal, so that nothing can escape.

Your matter will be placed on a well flattened, slightly warm bed; you will leave it there for a philosophical month; you will maintain the heat even, as long as the sweat of matter sublimates, until it no longer sweats, that nothing goes up, nothing goes down, it begins to rot, to suffocate, to coagulate and to settle, by result of the constancy of fire.

No more smoky aerial substance will rise and our Mercury will remain at the bottom, dry, stripped of its moisture, rotten, coagulated, changed into a black earth, which is called Black head of the crow, dry earthy element. When you have done this, you will have accomplished the true sublimation of the Philosophers, during which you have traversed all the aforementioned degrees; sublimation of Mercury, distillation, coagulation, putrefaction, calcination, fixation, in a single vessel and a single furnace, as has been said.

Indeed, when our stone is in its vessel, and it rises, it is then said that there a sublimation or ascension. But when it then sinks to the bottom, we say there a distillation or precipitation.

Then when after sublimation and distillation, our Stone begins to rot and coagulate, it is putrefaction and coagulation;

finally when it calcines and fixes by deprivation of its aqueous radical humidity, it is calcination and fixation;

all this is done by the single act of heating, in a single furnace, in a single vessel, as has been said.

This sublimation constitutes a true separation of the elements, according to the philosophers: "The work of our stone consists only of the separation and conjunction of the elements; for in our sublimation the cold and humid aqueous element is changed into a dry and hot earthy element."

It follows that the separation of the elements of our stone, is not vulgar, but philosophical; our only very perfect sublimation is indeed enough to separate the elements; in our stone there is only the form of two elements, water and earth, which virtually contain the other two.

The earth contains virtually Fire, because of its dryness; the water virtually encloses Air because of its humidity.

It is therefore quite obvious that if our Stone has in it only the form of two elements, it contains virtually all four of them.

So a Philosopher said:

> "There is no separation of the four elements in our Stone as fools think. Our nature contains an arcana very hidden whose strength and power we see, the earth and water. It contains two other elements, air and fire, but they are neither visible nor tangible, we cannot represent them, nothing detects them, we ignore their power, which only manifests itself in the two other elements, earth and water, when the fire changes colors during cooking. Behold, by the grace of God, you have the second component of the philosopher's stone, which is the Black Earth. "

Raven's head, mother, heart, root of other colors. From this earth as from a trunk, all the rest is born. This earthy, dry element has received the books of the philosophers a large number of names, we still calls it filthy Laton, black residue, Brass of the philosophers, Nummus, Black Sulphur, male, spouse, etc.

Despite this infinite variety of names, it is never one and the same thing, drawn from a single material.

As a result of this moisture deprivation, caused by philosophical sublimation, the volatile became fixed, the soft hard, the watery became earthy, according to Geber.

It is the metamorphosis of nature, the change of water on fire, according to Peat.

It is again the change of cold and humid constitutions into bilious, dry constitutions, according to the doctors.

Aristotle says that the spirit took on a body, and Alphidius that the liquid became viscous.

The occult has become manifest, says Rudianus in the Lipre of the Three Words.

We understand now the philosophers when they say:

> 'Our Great Work is nothing but a permutation of natures, an evolution of the elements. It is quite evident that by this deprivation of humidity we make the stone dry, the bird becomes fixed, the spirit becomes corporeal, the liquid becomes solid, the fire changes into water, the air into earth. We have thus changed the true natures according to a certain order, we have made the four elements turn in a circle, we have permuted their natures. '

May God be eternally blessed!
Amen.

Let us now pass with the permission of God to the second operation which is the bleaching of our pure land.

So take two parts of fixed earth or Head of crow; grind it subtly and carefully into one excessively clean mortar, add a part of the philosophical Water that you know (it is the water that you have put aside).

Do your best to unite them, soaking little by little little water the earth dries up, until it has sealed his thirst; ground and mixed so well, that the union of body, soul and water is perfect and intimate.

This done, you will put everything in a hermetically sealed matrass so that nothing escapes, and you will place him on his little smooth bed, lukewarm, always warm so that while sweating he rids his entrails of the liquid he has drunk. You will leave it there for eight days, until the earth turns white.
You will then take the Stone, you will pulverize it, you will soak it again with Virginal Milk, stirring, until she quenches her thirst; you will put it back in the vial on its little warm bed so that it dries up sweating, as above.

You will repeat this operation four times following the same order: soaking the earth with water until perfect union, desiccation, calcination. You will have cooked enough Earth of our very precious stone.

By following this order; cooking, pulverization, imbibition with water, desiccation, calcination, you have sufficiently purified the Head of raven, the black and fetid earth, you brought it to whiteness by the power of fire, heat and whitening water.

Collect your white earth and put it carefully aside, because it is a precious good, it is the white foliated Earth. White Sulphur, White Magnesia, etc.
Morien speaks of her when he says...
"Putrify this earth with its water, so that it may be purified and with God's help you will complete the Magisterium. "

Hermes likewise says that:

" the Azoth washes the Latin and removes all impurities."
In this last operation we have reproduced the true conjunction of the elements, because water has united with earth, air with fire. It is the union of man and woman, of male and female, of gold and silver, dry sulfur and impure celestial water.
There's was also resurrection of dead bodies.

This is why the philosopher said: "Let those who do not know how to kill and resuscitate abandon art"

and elsewhere: "Those who know killing and resurrecting will profit in our science. He will be the Prince of Art who will be able to do these two things."

Another philosopher said,

"Our dry earth will bear no fruit unless it is deeply soaked with its Rainwater. Our dry earth has a great thirst, when she starts to drink, she drinks until the end. "

Another said:

"Our Earth drinks the fertilizing water it has been waiting for, it quenches its thirst, and then it produces hundreds of fruits."

We find many other similar passages in the books of the philosophers, but they are in the form of a parable, so that the wicked cannot hear them.

By the grace of God, you now possess our foliated white Earth ready to undergo fermentation, which will give it breath.

Also the Philosopher said: "Whiten the black earth before adding the ferment to it. Another said:

"Sow your gold in the White Foliage....and it will yield you a hundredfold fruit. Glory to God. Amen."

Let's move on to the third operation which is the fermentation of the White Earth. We must animate the dead body and resuscitate it, to multiply its power to infinity, and make it pass to the state of perfect white Elixir which changes Mercury into the perfect and true Moon.

Note that the ferment can only penetrate the dead body through the intermediary of the water which makes the marriage and serves as a link between the white earth and the ferment.

This is why in any fermentation, it is necessary to note the weight of everything.

If therefore you want to ferment the white foliated earth to change it into a white elixir containing an excess of tincture, you must take three parts of white earth or foliated dead body, two parts of the brandy that you have put in reserve and a part and a half of ferment.

Prepares the ferment in such a way that it is reduced to a thin and fixed white lime if you want to make the white elixir.

If you want to make the red elixir, use very yellow gold lime, prepared according to art. There are no other ferments than these.

The leaven of silver is silver, the leaven of gold is gold, so look no further.

The reason is that these two bodies are luminous, they contain rays dazzling which communicate to other bodies the true redness and whiteness.

They are similar in nature to that of the purest Sulfur of matter, of the species of stones.

Extract therefore each species from its species, every kind of its kind.
The white work has the aim to whiten, the red work to blush.
Do not mix the two Works, otherwise you will do nothing Good.

All the Philosophers say that our Stone is made up of three things: the body, the spirit and the soul.

Gold, the leafy white earth is the body,
the ferment is the soul which gives it life,
the intermediate water is the spirit.

Unite these three things into one through marriage,
grinding them well on a clean stone,
so as to unite them in their most infinite particles,
to form one confused chaos.

(W) Stir the mixture again

When you have made one body of all, you will put it gently in a special vial, which you will place on its warm bed, so that the mixture coagulates, settles and turns white.

You will take this blessed white stone, you will grind it subtly on a clean stone,
you will soak it with a third of its weight of water to lower his thirst.

You will then put her back in the clear and clean vial on her warm bed, so that it begins to sweat, to give up its water

and finally you will let its entrails dry up.

Repeat several times until you have prepared by this process our very excellent white stone, fixed, which penetrates the smallest parts of the body very quickly, flowing like fixed water when put on the fire, turning imperfect bodies into real silver, comparable in all respects to natural silver.

 Note that if you repeat all these operations several times in the same order: dissolve, coagulate, grind, cook, your Medicine will be all the better, its excellence will increase more and more.

The more you will work your Stone to increase its virtue, and the more yield you will have when you make the projection on imperfect bodies. So that after one operation a part of the Elixir changes one hundred parts of any body into Moon, after two operations one thousand, after three ten thousand, after four hundred thousand, after five a million, after six operations thousands of thousands and so on ad infinitum.

Also the adepts all praise the great maxim philosophers on the perseverance to repeat this operation.

If an imbibition had sufficed, they would not have talked so much on this subject. Thanks be given farewell. Amen.

If you wish to change this glorious Stone, this King
white which transmutes and tints Mercury and all bodies
imperfect in true Moon,

 if you wish,
I say, to change it in red Stone which transmutes and dyes the Mercury, the Moon and the other metals in true Sun, operates thus.

Take the White Stone and divide it into two parts;
you will increase one to the state of white elixir with its Water white,
as was said above, so that you will have it indefinitely.

You will put the other in the new bed of the philosophers, neat, clean, transparent, spherical, and you will place the whole thing in the digestion furnace.

You increase the fire until by its force and power matter is changed into a very red stone, which the Philosophers call Blood, purple gold. Coral red. Red sulfur.

When you see this color such that the red is as bright as charred dry crocus,
So take the King Merrily, put it precisely next to.

If you want to change it into a tincture of the very powerful Red Elixir, transmuting and coloring the Mercury, the Moon and any other imperfect metal into the very true Sun, ferment three parts of it with one part and a half of very pure gold in the lime state held and well yellow, and two parts of solidified water.

Make a perfect mixture according to the rules of the Art, until no more distinguish nothing from the components. Put back in the vial over a fire that matures, to give it perfection.

As soon as the true Red Bloodstone appears, you will gradually add Solid Water. You will gradually increase the fire of digestion. You increase its perfection by repeating the operation.

It is necessary each time to add solid Water (which you have kept),
which suits its nature; it multiplies its power to infinity, without changing anything in its essence.

One part of perfect Elixir in the first degree projected on one hundred parts of Mercury (washed with vinegar and salt, as you must know) placed in a crucible slowly, until fumes appear, immediately transmutes them into real Sun better than the natural.

Likewise by replacing Mercury with the Moon.
For each degree of perfection in addition to the Elixir,
it is the same as for the White Elixir, until
let it dye finally in Sun
 infinite quantities of Mercury and Moon.
You now have a precious arcane, an infinite treasure.

This is why philosophers say:

"Our Stone has three colors, it is black at the beginning, white in the middle, red at the end. "

A philosopher said:

"Heat acting first on the humid engenders blackness, its action on the dry engenders whiteness and on the whiteness engenders redness. For whiteness is nothing other than the complete deprivation of darkness. White, strongly condensed by the force of fire, engenders red. "

- "All you seekers who work the Art, said another sage, when you see the white appear in the vessel, know that the red is hidden in this white. You have to extract it and for that heat strongly until the appearance of red. "

Now give thanks to God sublime and glorious
Sovereign of Nature, who created this substance and
gave it a property that is not found in any other body.

(R) Allow the solution to settle

It is she who, set on fire, engages the fight with this one and resists him valiantly. All the other bodies flee or are exterminated by fire.

Collect my words, note how much they contain of mysteries, because in this short treatise I have collected and explained what is the most secret in Alchemy; everything there is said simply and clearly, I have not omitted anything, everything is there briefly indicated, and I take God to witness that in the books of the Philosophers, one cannot find nothing better than what I told you.

Also I beg you, do not entrust this treatise to anyone, do not let it fall into impious hands, for it contains the secrets of philosophers of all ages. A tea quantity of precious pearls should not be thrown to swine and the unworthy.

If, however, this should happen, I then pray to Almighty God that you never manage to complete this divine Work.

        Blessed be God, one in three persons.
        Amen.

GLOSSARY

(Aigle Volant) Flying eagle. — Mercury of the philosophers.

Alphadius. — Greek philosopher; manuscript 65 14 of the national library: Liber meteorum, east of Alphadius. Despite its title, it is a treatise on alchemy.

Aludel. — Sublimation apparatus, consisting of a glazed earthenware vase, surmounted by a glass capital intended to receive the sublimation.
Soul. "It is the volatile part of the stone; more especially this word designates the ferment.

Aristotle. —Disciple of Avicenna, he must not be "fused with the Greek philosopher, Alexander's tutor.
Works: De perfecto magisterio. —De practica Lapidis.
spagyric art. — Synonym of Alchemy. Astanus. "Perhaps he is the same as Ostanes. There are several manuscripts of the latter in libraries.

Star. — It is the essential principle of metals capable of changing bodies into its own nature (Paracelsus).'

Avicenna. — Al Hussein Ebn Sina, born in Bokhara, disciple of Alpharabi, Arab alchemist, lived in the eleventh century.
Works: Canon medicinæ; Tractatulus alchemiae; Of conglutinalione lapidum.

Azoth. — Mercury of the Philosophers. Basil Valentine & makes a treatise on Azoth.

Barsen or Basen. — Alchemist quoted in the Peat.

Red lime. — Matter from Stone to Red.

Cocober. — Return to the retort the liquid that has distilled. Body. — Fixed part of the stone.

Crocus. — Matter from Stone to Red. Means :also oxidizes.

Degrees. — The first degree of fire corresponds to 50 degrees centigrade, the second to the boiling of water, the third to the fusion of tin, the fourth to the boiling of mercury.

Water, holy water, metallic water, cloud water, brandy — Mercury of the philosophers.

Entelechy: essential and perfect form of a thing.

Ash fire. — Sand bath.

Geber. — Djafar al Soli, Arab hermetic philosopher, deprived in the ninth century. He is the most famous of the Arab alchemists. Works: Sum of perfection. — Will.

(Gros.) Fat. — Old measure of weight: 3 gr., 90. Un large is 72 grains.

Hermes. — Egyptian Thaut, the father of chemistry. Or" turns: Table of Emerald. — The Seven Chapters.

Hyle. — Matter of stone, Mercury of philosophers phes.

Latonou brass. — Mercury of the philosophers before putrefaction, that is to say before darkness.

Lion.— Green lion, green vitriol. Red lion, hypo-nitrogen gas.
(Lot) Batch. — German measure of weight, equivalent to a

half ounce.

Moon. — Silver, or ordinary mercury, or matter to white. Magisterium. — Synonym of philosopher's stone and Great Work.

Mercury of the philosophers. — Raw material of the rock.

Metals. — The alchemists recognize only seven, to which they attribute the names of the planets. Gold or Sun, Silver or Moon, Mercury, Lead or Saturn, Tin or Jupiter, Copper or Venus, Iron or Mars.

Medicine. — Synonym of elixir.

Microcosm. - Or small world, it's the man, bam
opposition to the macrocosm, which is the universe. The philosophers still mean by microcosm their magisterium.

Morian. — Disciple of Adfar, hermetic philosopher Roman.
Works:
    On the transmutation of metals -
    Dialogue between King Calid and the philosopher Morien.
    Ounce. Old measure of weight is equivalent to 3 1 gr. 2 5

.
Hermes bird. — Mercury of the philosophers.

Pelican. — Circulatory vase: it consists of a
belly surmounted by a capital, from which start two
tubes which enter the rumen laterally, so that the distilling liquid constantly falls back into the rumen.

Phoenix. — Fabulous bird with red plumage; red stone emblem.

Physician. - Doctor. This word is still used in England in this sense (physician).
Chicken of Hermogenes. — Material from stone to white.

Rhases. — Persian hermetic philosopher, lived in the tenth century. His works exist in manuscript in Latin translation at the national library:

Lumen luminum.

Liber per/ecli magisterii.

From aluminum and salibus. Sulfur.

Second principle of metals. also means the material of stone. Vivid sulfur or red sulfur, matter from stone to red.

Sublimation. — In the philosophical sense, means purification.

Raven's head. — It is matter during putrefaction.

Theleme. — Perfection.

Peat of the philosophers. — Turba philosophorum, the best known and the most common of the ancient treatises on Alchemy. Attributed to the philosopher Aristeas.

Belly of the horse. — Hot horse manure, provides a constant temperature.

Vinegar, white, very sour, philosophers. — Mercury of the philosophers.

Vitriol. — Green vitriol or Roman vitriol is all one, means green rosacea. Blue vitriol or

(K) It receives the force of things superior and inferior

Hungary: blue couperose.

Note. — Rudianus, Franciscus, and Mechardus are completely unknown. These are probably Greek alchemists whose works are lost.

(X) Allow the mixture to cool

Milton Keynes UK
Ingram Content Group UK Ltd.
UKHW052000120923
428564UK00007B/102

9 781312 355255